The Telomere

The Telomere

David Kipling

Medical Research Council, Human Genetics Unit,
Western General Hospital, Edinburgh

Oxford   New York   Tokyo
OXFORD UNIVERSITY PRESS

This book has been printed digitally in order to ensure its continuing availability

OXFORD
UNIVERSITY PRESS

Great Clarendon Street, Oxford OX2 6DP

Oxford University Press is a department of the University of Oxford.
It furthers the University's objective of excellence in research, scholarship,
and education by publishing worldwide in

Oxford New York

Auckland Bangkok Buenos Aires Cape Town Chennai
Dar es Salaam Delhi Hong Kong Istanbul Karachi Kolkata
Kuala Lumpur Madrid Melbourne Mexico City Mumbai Nairobi
São Paulo Shanghai Singapore Taipei Tokyo Toronto
with an associated company in Berlin

Oxford is a registered trade mark of Oxford University Press
in the UK and in certain other countries

Published in the United States
by Oxford University Press Inc., New York

A catalogue record for this book is available from the British Library

Library of Congress Cataloging in Publication Data
Kipling, David.
The telomere / David Kipling.
Includes bibliographical references and index.
1. Telomere. I. Title.
QH600.3.K547   1995   574.87'332—dc20   94-33104

ISBN 0-19-963600-1 (Pbk)

Preface

The aim of any review is to provide a bridge to the primary literature, to enable the reader to progress to experimental papers, read them in context, and understand the question being investigated. The study of telomeres, the specialized nucleoprotein structures that terminate linear chromosomes, is a particularly exciting and diverse field. This book aims to introduce the concepts and central questions along with the body of experimental data. I hope it will be an accessible introduction to a field full of fascinating and unique phenomena for the advanced undergraduate. The main target is the molecular biologist new to telomeres but my hope is that the treatment of such a large reference base will continue to make it of use to those more familiar with the field.

Acknowledgements

Many friends and colleagues have helped with the preparation of this book by reading and commenting on various chapters (in some heroic cases the entire manuscript) and by providing preprints. This book would have been an impossible feat for one person without their help; to Robin Allshire, Rosemary Bayne, Judy Berman, Harald Biessmann, Sarah Bowen, Kiki Broccoli, Howard Cooke, Titia de Lange, Susan Gasser, Carol Greider, Brenda Grimes, Calvin Harley, Shona Kerr, James Mason, Carol Newlon, Carolyn Price, David Shore, Bob Speed, and Helen Wilson I am deeply grateful. Whatever confusions, omissions, and errors remain, *mea culpa*. I would like to thank Sheila Mould, as this book would not have been possible without access to the excellent Human Genetics Unit library of which she is librarian. I am grateful to Judy Fantes for providing the *in situ* hybridization shown on the cover. My special thanks go to Sandy Bruce and Douglas Stuart for producing initial versions of most of the figures.

Contents

1

Introduction

This book begins with the central biological problem that telomeres circumvent, that of replicating the very end of a linear DNA molecule (for general reviews on telomeres see Szostak 1983; Blackburn and Szostak 1984; Zakian 1989; Zakian *et al*. 1990; Blackburn 1991*a,b*; Greider 1991; Henderson and Larson 1991; Biessmann and Mason 1992). Nature has devised a fascinating selection of mechanisms to do this (below) although much interest centres on a type of DNA polymerase called telomerase, whose biochemistry is detailed in Chapter 4. Many of the phenomena the current molecular approaches are trying to explain concern the behaviour of chromosome ends observed by microscopy (Chapter 2). It was the failure to obtain terminal deletions and inversions in X-irradiated *Drosophila* which first defined the natural end of a chromosome as having a special structure and function required for chromosome stability, the telomere (Muller 1938, 1940; reviewed by Muller and Herskowitz 1954). Cytological observations of the fate of broken ends in maize led McClintock in the 1940s to suggest that one such function was to protect the natural ends of chromosomes from recombination and fusion with other chromosomes. Telomeres may also be involved in the three-dimensional organization of the interphase nucleus, via interactions with the nuclear envelope, the nuclear matrix, and other telomeres. Organization and compartmentalization of the nucleus in three dimensions has important biological implications for processes such as replication, transcription, and chromosome segregation. An apparent organizational role for telomeres is seen in meiosis, where interactions with the nuclear envelope may play a role in facilitating the alignment and pairing of homologous chromosomes. This in turn has important consequences for how genetic maps are interpreted.

How the DNA sequences at telomeres are isolated and studied is described in Chapter 3. There are a number of distinct features of telomeric sequences that any model of telomere synthesis and maintenance must seek to explain. The short repeat sequences at telomeres can be synthesized by an enzyme called telomerase (Chapter 4). It is a fascinating enzyme in its own right, and its biochemistry is able to explain (or at least is consistent with) many features of telomeric DNA. The analysis of telomerase biochemistry is playing an important role in interpreting telomeric sequence data and understanding telomere behaviour. This behaviour can also be investigated by studying other *trans*-acting factors that interact with telomeres (Chapter 5).

The structure and function of telomeres detailed in previous chapters is then used to attempt to explain the role of telomeres in various examples of genome rearrangements, in particular chromosome breakage in species as diverse as humans and ciliates. Following this theme, the question is asked: what happens if telomeres are lost? In humans progressive telomere loss accompanies ageing and is also seen in some examples of neoplasia. Genome instability caused by telomere loss and the subsequent recombinogenic nature of the ends of the chromosomes may have important clinical consequences.

Telomeres, at least in yeast, have an altered chromatin structure able to influence both the transcription of adjacent genes, in a fashion similar to that seen with position effect variegation in *Drosophila*, and the replication timing of adjacent genomic regions. As explained in Chapter 8, this is best understood in yeast, but these results have important implications for telomere behaviour in other species.

The penultimate chapter details a rather strange tale. The concept of the telomere was first developed in *Drosophila*, following Muller's inability to obtain viable terminal deletions of *Drosophila* chromosomes. However, in a rather ironic twist it may turn out that *Drosophila* is one organism where a special terminal DNA sequence is not required to stabilize the chromosome. Viable terminal deletions in *Drosophila* have now been created, and this species may also overcome the end-replication problem in a novel fashion. To conclude, Chapter 10 puts the structure and behaviour of telomeres into the context of human genome analysis and genetic disease.

The end-replication problem

This is the central biological problem that telomeres are designed to overcome or compensate for, and is illustrated in Fig. 1.1. It stems from the fact that all known DNA polymerase enzymes require a 3'-OH group on which to add the next nucleotide (Wang 1991) and that synthesis cannot therefore start *de novo*. For normal lagging strand replication, short RNA primers are made by primase, an enzyme which does not require a free hydroxyl group and is able therefore to start synthesis *de novo*. These RNA primers provide the necessary 3'-OH groups for DNA polymerase, which takes over and extends the chains to synthesize Okazaki fragments. The subsequent removal of these RNA primers results in an inability to synthesize lagging strand sequence complementary to a small region at the very end of the chromosome (Fig. 1.1). This 'primer gap' can also result from Okazaki fragments not starting at the very end of the molecule, and will be at least as large as the RNA primer lost. Every round of synthesis will result in a small loss of terminal sequence, and would inexorably remove

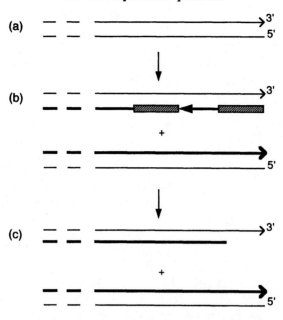

Fig. 1.1 The end-replication problem. The very end of the chromosome is replicated by a fork moving from the left. Leading strand synthesis results in complete duplication of the upper strand. Lagging strand synthesis of the lower strand is initiated by RNA primers (hatched boxes). Their removal is followed by Okasaki fragment extension to fill the gaps. However, a region at the very end of the molecule cannot be synthesized.

sequence from the chromosome (Olovnikov 1971, 1973; Watson 1972).

Why do DNA polymerases have this substrate requirement? One consequence is that it enables DNA polymerases to contribute to error correction (Wang 1991; Kunkel 1992). A single mispaired nucleotide at the 3' end of the chain being extended inhibits subsequent chain extension. For example, calf thymus DNA polymerase α extends mispaired nucleotides 10^3–10^6 times more slowly than a correctly base-paired terminus. A 3' exonuclease removes the mispaired base and chain extension continues. Such 'proof-reading' exonucleases typically have a preference for mispaired nucleotides. However, the ability to proof-read is greatly assisted by DNA polymerase halting synthesis at mispaired bases, and although this proof-reading and editing is not perfect it does result in a much reduced rate of incorrect base incorporations relative to that which occurs with enzymes such as RNA polymerases which do not proof-read (reviewed by Echols and Goodman 1991).

Potential solutions to the end-replication problem

Linear DNA molecules are widespread, and nature has evolved numerous ways to overcome or compensate for the end-replication problem.

Terminal hairpins

Some of the earliest models invoked terminal hairpins to overcome the end-replication problem (reviewed by Szostak 1983; Blackburn and Szostak 1984; Zakian 1989). Although there are a number of variants, two of which are illustrated in Fig. 1.2, all centre on the termination of the DNA duplex in a perfect hairpin loop, formed by a combination of a palindromic sequence and a covalent linkage between the two DNA strands.

Hairpin termini are found for a wide range of linear DNA species, including some bacterial plasmids (Barbour and Garon 1987) and the mitochondrial DNA of some species (Pritchard and Cummings 1981; Martin 1991). Their existence alone does not constitute proof that the hairpins are involved in end-replication, as they might be the product of some unconnected event such as recombination. However, there is evidence that they are used for telomere replication in a number of animal viruses, in particular poxviruses such as the vaccinia virus (reviewed by DeLange and McFadden 1990). However, this mode of telomere replication has yet to be observed for eukaryotic chromosomal telomeres.

One model of hairpin terminus replication is illustrated on the right of Fig. 1.2. The presence of the covalently closed hairpin termini means that the molecule is, in effect, a single-stranded circle. Because of this, conventional DNA replication can replicate the entire molecule. Following this, a site-specific endonuclease nicks the DNA to reveal two single-stranded termini. As these are palindromic they can fold back, anneal, and ligate to reform the initial structure. Thus, by a combination of an unusual form for the DNA duplex and conventional replication enzymes the termini of the molecule are replicated.

Another model invokes branch migration following DNA replication (leftmost pathway in Fig 1.2). This creates a cruciform structure which is nicked; the ends separate and the strands ligate to reform the hairpin termini. There are a number of well-characterized endonucleases that will nick cruciform structures in just this fashion, such as T4 endonuclease VII (reviewed by West 1992). In this case resolution can be thought of as essentially a site-specific recombination reaction using a 'Holliday resolvase' to produce the desired nicking event.

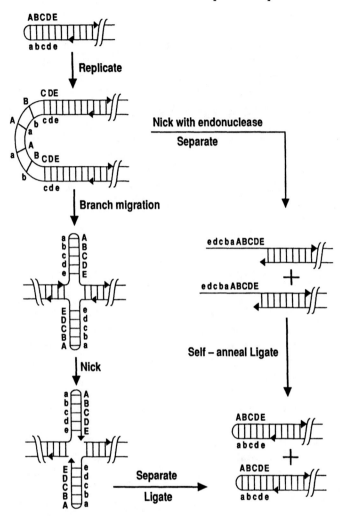

Fig. 1.2 Telomere replication using hairpin termini. A linear DNA duplex is shown terminating in a palindrome, with the two strands covalently linked. Following conventional DNA replication, branch migration could occur to form a cruciform structure which can be nicked by a Holliday resolvase (left). Alternatively, a site-specific endonuclease could cleave and the single-stranded ends anneal back on themselves (right). A ligation reaction reforms the initial hairpin.

Recombination

The essence of this model is the use of some form of unequal gene conversion event to convert a short telomere to a longer one, producing a net synthesis of telomeric sequences. The model illustrated in Fig. 1.3 shows a gene conversion event in which the 3′ end of a telomere invades an internal site at another telomere and is then elongated by DNA polymerase on this template.

There has been debate regarding to what extent recombination plays a role in yeast telomere replication. Certainly recombination between non-yeast terminal sequences does occur when introduced into yeast. However, there remains little direct evidence that recombination plays any role in the normal maintenance of yeast telomeres. This topic is dealt with in greater depth in Chapter 6.

Protein-primed replication

Proteins are found covalently linked to the 5′ ends of many double-stranded linear DNAs. Some of these proteins may be involved in biological activities such as the encapsulation of viral genomes, but a number have a clearly defined role in priming rounds of DNA replication. In this mode of replication, the primer is not the free 3′-OH group of an RNA or DNA molecule, but the -OH group of a serine, threonine, or tyrosine residue of this terminal protein (TP).

Protein-primed replication initiation has been detailed most extensively in systems able to initiate replication *in vitro* with purified proteins, such as those for adenovirus and bacteriophage Φ29 (Challberg and Kelly 1989; Salas 1991). Protein-primed replication has also been demonstrated for bacteriophage PRD1 and Cp-1, and there are TPs on a number of other linear DNAs and plasmids which therefore may also replicate in this fashion.

A generalized scheme for protein-primed replication is shown in Fig. 1.4. It commences with the interaction of initiation proteins with *ori*, the origin of replication. These cause the DNA helix to unwind, allowing access of other factors to the DNA template. A complex then forms at *ori* containing TP and DNA polymerase. A covalent bond is formed between TP and a deoxymononucleotide (dCMP in the case of adenovirus replication). The bond is with the -OH group of either a serine, threonine, or tyrosine residue on the TP (Ser580 for adenovirus). DNA polymerase uses this TP-dNMP as the primer for replication and proceeds to catalyse chain elongation. The end result is a full length DNA molecule with the TP covalently attached to the 5′ end.

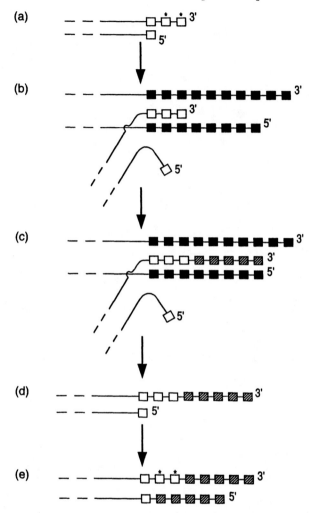

Fig. 1.3 Telomere replication by recombination. The region of the G-rich upper strand that cannot be copied by conventional DNA replication is indicated by asterisks in (a). In (b), this strand has invaded another telomere and base-paired with it. Repair synthesis, (c), results in a net addition of sequence to this strand. Conventional lagging-strand DNA replication now has sufficient sequence distal to the asterisked region to be able to prime and synthesize the complementary sequence.

Introduction

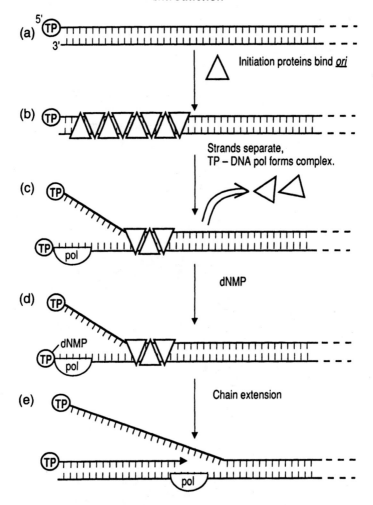

Fig. 1.4 Generalized model for protein-primed DNA replication. (a) The initial
molecule is shown with a terminal protein (TP) covalently attached to its 5' terminus. (b) Initiation proteins bind the origin of replication, *ori*. (c) Strand separation occurs, allowing formation of a complex at *ori* containing DNA polymerase
(pol) and TP. (d) A nucleotide monophosphate (dNMP) becomes covalently
attached to TP and acts as the first base of synthesis. (e) TP-dNMP is used as the
primer for chain elongation by DNA polymerase.

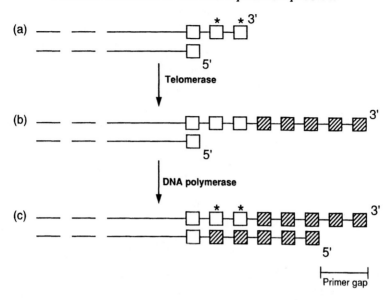

Fig. 1.5 Telomere replication by telomerase. The region of the G-rich upper strand that cannot be copied by conventional DNA replication is indicated by asterisks in (a). Telomerase action adds sequence to this strand in (b). Conventional lagging-strand DNA replication now has sufficient sequence distal to the asterisked region to be able to prime and synthesize its complementary sequence. As in Fig. 1.3, although net synthesis has occured a new 'primer gap' has been created and the upper strand terminates in a single-stranded region.

Telomerase

The sequence that is lost because of incomplete replication is that complementary to the terminus of the strand running 5′ to 3′ towards the end of the chromosome (the upper strand in Fig. 1.5). In a number of species a special enzymatic activity has been identified called *telomere terminal transferase*, or more commonly *telomerase*. This enzyme adds telomeric sequence onto this upper strand, with conventional lagging strand DNA synthesis replicating the other strand. Together, these activities can cause a net increase in the sequence on both strands to compensate for any sequence lost through the end-replication problem. The predicted result is a treadmilling of terminal repeat sequences, with a balance between the continual processes of loss and addition. Telomerase is a main theme of this book and is described in depth in subsequent chapters.

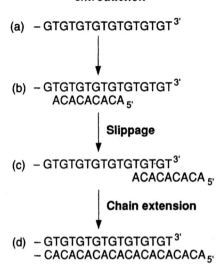

(a) – GTGTGTGTGTGTGTGTGT $^{3'}$

(b) – GTGTGTGTGTGTGTGTGT $^{3'}$
ACACACACA $_{5'}$

Slippage

(c) – GTGTGTGTGTGTGTGT $^{3'}$
ACACACACA $_{5'}$

Chain extension

(d) – GTGTGTGTGTGTGTGT $^{3'}$
– CACACACACACACACACA $_{5'}$

Fig. 1.6 Simple sequence slippage during DNA replication. Because of the repetitive nature of the template, newly replicated strands can 'slip' during synthesis. Slippage might result in strands partially slipping off the end, or regions looping out. The end result is net synthesis; a greater length of complementary strand can be made than that simply directed by the template.

Unusual aspects of DNA sequence metabolism

Telomeres are often composed of repeated short, simple sequences, such as $(TTAGGG)_n$ (Chapter 3). There is some evidence to suggest that such simple sequences may have unusual properties during their replication. For example, DNA polymerase may be able to 'slip' on them whilst using them as templates for replication. DNA polymerase slippage has been shown *in vitro* (Schlötterer and Tautz 1992) for sequences such as $(CA)_n$. As illustrated in Fig. 1.6, such slippage of short nascent DNA fragments can result in synthesis of complementary strands longer than the template. The similarity between sequences which will undergo slippage *in vitro* and telomeric sequences may point to a mechanism to cause net sequence addition at telomeres. However, most telomeric sequences have not been tested for slippage *in vitro* and this mechanism remains very speculative.

Slippage is also of significance to the stability of non-terminally located simple sequences. Internal $(GT)_n$ tracts are highly unstable in yeast, showing frequent (>1 per 10^4 divisions) loss or addition of GT repeat units (Henderson and Petes 1992), possibly as a result of polymerase slippage. This appears to be an intrinsic feature of this type of sequence, which is similarly unstable in prokaryotes. The significance of slippage synthesis to

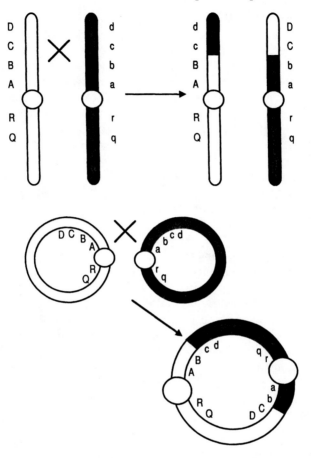

Fig. 1.7 Ring chromosome behaviour. Meiotic recombination between chromosomes, or sister chromatid exchange in somatic cells, can produce a dicentric chromosome from two ring chromosomes (lower). Such crossovers do not cause this problem for linear chromosomes (upper).

both telomeric and non-telomeric sequence hypervariability and instability in the human genome is discussed in Chapter 10.

Circular genomes

The simplest way to avoid the end-replication problem is not to have a linear genome. This is the solution adopted by *Escherichia coli*, and even the apparently linear bacteriophage λ converts itself to a covalently closed circle at the appropriate point in its life cycle by annealing its *cos* sticky ends. However, recombination can pose a major problem for circular genomes. For example, in *E. coli* newly replicated chromosomes can become covalently linked by homologous recombination, with resolution to monomers relying on site-specific recombination by XerC and XerC resolvase proteins acting at the *dif* locus (Løbner-Olesen and Kuempel 1992). In a similar fashion recombination between ring chromosomes in eukaryotes can lead to a chromosome with two centromeres (Fig. 1.7). However, it does not seem that eukaryotes possess a specific mechanism to resolve such dicentric chromosomes, which can then be broken in mitosis if the two centromeres are pulled to separate poles. Recombination naturally occurs in meiosis, and sister chromatid exchange can also lead to dicentrics being formed during somatic growth. In humans it has been reported that patients with circular ('ring') chromosomes that do not appear to have any genetic material deleted still manifest clinical symptoms of what is termed 'ring syndrome'. It has been suggested that these symptoms result from the instability of the ring chromosome during somatic growth, leading to chromosome rearrangements and aneuploidy in the somatic tissue (Pezzolo *et al.* 1993; Sawyer *et al.* 1993). Thus the preponderance of linear chromosomes in eukaryotes may reflect their sensitivity to the aneuploidy (especially at the level of the whole organism) that can be the result of a circular genome. Instead of devising a specific mechanism to resolve dicentrics, the problem is avoided altogether by having linear chromosomes capped by telomeres.

References

Reviews

Telomeres (general)

Biessmann, H. and Mason, J. M. (1992). Genetics and molecular biology of telomeres. *Adv. Genet.*, **30**, 185-249.
Blackburn, E. H. (1991*a*). Structure and function of telomeres. *Nature*, **350**, 569-73.
Blackburn, E. H. (1991*b*). Telomeres. *Trends Biochem. Sci.*, **16**, 378-81.
Blackburn, E. H. and Szostak, J. W. (1984). The molecular structure of centromeres and telomeres. *Annu. Rev. Biochem.*, **53**, 163-94.
Greider, C. W. (1991). Telomeres. *Curr. Opin. Cell Biol.*, **3**, 444-51.
Henderson, E. R. and Larson, D. D. (1991). Telomeres—what's new at the end? *Curr. Opin. Genet. Dev.*, **1**, 538-43.
Szostak, J. W. (1983). Replication and resolution of telomeres in yeast. *Cold Spring Harbor Symp. Quant. Biol.*, **47**, 1187-94.
Zakian, V. A. (1989). Structure and function of telomeres. *Annu. Rev. Genet.*, **23**, 579-604.
Zakian, V. A., Runge, K., and Wang, S.-S. (1990). How does the end begin? Formation and maintenance of telomeres in ciliates and yeast. *Trends Genet.*, **6**, 12-16.

Other topics

Challberg, M. D. and Kelly, T. J. (1989). Animal virus DNA replication. *Annu. Rev. Biochem.*, **58**, 671–717.

DeLange, A. M. and McFadden, G. (1990). The role of telomeres in poxvirus DNA replication. *Curr. Topics Microbiol. Immunol.*, **163**, 71–92.

Echols, H. and Goodman, M. F. (1991). Fidelity mechanisms in DNA replication. *Annu. Rev. Biochem.*, **60**, 477–511.

Kunkel, T. A. (1992). Biological asymmetries and the fidelity of eukaryotic DNA replication. *Bioessays*, **14**, 303–8.

Salas, M. (1991). Protein-priming of DNA replication. *Annu. Rev. Biochem.*, **60**, 39–71.

Wang, T. S.-F. (1991). Eukaryotic DNA polymerases. *Annu. Rev. Biochem.*, **60**, 513–52.

West, S. C. (1992). Enzymes and molecular mechanisms of genetic recombination. *Annu. Rev. Biochem.*, **61**, 603–40.

Primary papers

Barbour, A. G. and Garon, C. F. (1987). Linear plasmids of the bacterium *Borrelia burgdorferi* have covalently closed ends. *Science*, **237**, 409–411.

Henderson, S. T. and Petes, T. D. (1992). Instability of simple sequence DNA in *Saccharomyces cerevisiae*. *Mol. Cell. Biol.*, **12**, 2749–57.

Løbner-Olesen, A. and Kuempel, P. L. (1992). Chromosome partitioning in *Escherichia coli*. *J. Bacteriol.*, **174**, 7883–9.

Martin, F. N. (1991). Linear mitochondrial molecules and intraspecific mitochondrial genome stability in a species of *Pythium*. *Genome*, **34**, 156–62.

Muller, H. J. (1938). The remaking of chromosomes. *Collect. Net*, **8**, 182–95.

Muller, H. J. (1940). An analysis of the process of structural change in the chromosomes of *Drosophila*. *J. Genet.*, **40**, 1–66.

Muller, H. J. and Herskowitz, I. H. (1954). Concerning the healing of chromosome ends produced by breakage in *Drosophila melanogaster*. *Am. Nat.*, **88**, 177–208.

Olovnikov, A. M. (1971). Principle of marginotomy in template synthesis of polynucleotides. (In Russian) *Dokl. Akad. Nauk S.S.S.R.*, **201**, 1496–9.

Olovnikov, A. M. (1973). A theory of marginotomy. *J. Theor. Biol.*, **41**, 181–90.

Pezzolo, A., Gimelli, G., Cohen, A., Lavaggetto, A., Romano, C., Fogu, G., and Zuffardi, O. (1993). Presence of telomeric and subtelomeric sequences at the fusion points of ring chromosomes indicates that the ring syndrome is caused by ring instability. *Human Genet.*, **92**, 23–7.

Pritchard, A. E. and Cummings, D. J. (1981). Replication of linear mitochondrial DNA from *Paramecium*: sequence and structure of the initiation-end crosslink. *Proc. Natl. Acad. Sci. USA*, **78**, 7341–5.

Sawyer, J. R., Rowe, R. A., Hassed, S. J., and Cunniff, C. (1993). High-resolution cytogenetic characterization of telomeric associations in ring chromosome 19. *Human Genet.*, **91**, 42–4.

Schlötterer, C. and Tautz, D. (1992). Slippage synthesis of simple sequence DNA. *Nucleic Acids Res.*, **20**, 211–15.

Watson, J. D. (1972). Origin of concatameric T7 DNA. *Nature New Biol.*, **239**, 197–201.

2

The cytology of the telomere

The concept that the natural end of a chromosome has a specialized structure came from a combination of genetic and cytological observations in the 1930s. This chapter reviews these early classical studies as well as cytological observations of interactions between telomeres, and between telomeres and the nuclear envelope. These interactions may have a role in organizing the genome within the interphase nucleus, and in facilitating the pairing of homologous chromosomes in meiosis. The first part of this chapter concentrates on organization in the mitotic nucleus, in which telomeres have at best a tentative role. Subsequently the behaviour of telomeres in meiosis is considered, where there is stronger evidence for telomeres having a functional role.

The telomere defined

Evidence for telomeres first came from light microscopic studies in *Drosophila* (Muller 1938, 1940; reviewed by Muller and Herskowitz 1954). It was observed that terminal deletions and inversions were not recovered following X-irradiation, whereas interstitial deletions and inversions were readily obtained. That is, rearrangements which removed the natural end of the chromosome and replaced it by a genomic region which is normally internal on the chromosome were not found. This suggested that the natural end of a linear chromosome has a specialized structure which causes it to behave in a way that is distinct from a free end. This hypothetical structure was termed the telomere. These classical studies have been confirmed in wild-type *Drosophila* (Roberts 1974, 1975) although it should be noted that in *mu-2* flies terminal deletions can be recovered at a reasonable frequency; how this apparent contradiction is reconciled with the functional definition of a telomere is the subject of Chapter 9.

Why were terminal deletions not recovered? One possible explanation came from studies of the fate of broken chromosomes in maize (McClintock 1938, 1939, 1941). Broken chromosomes were produced by the action of spindle forces on dicentric chromosomes; during chromosome segregation the two centromeres can be pulled to opposing poles thus breaking the chromosome. It was observed that the end of such a broken chromosome was very reactive and underwent aberrant recombination and fusion reactions with other chromosomes. This in turn could create a new dicentric chromosome, which would itself then undergo breakage and recombination reactions. Such breakage–fusion–bridge cycles were characteristic of the ends of broken chromosomes, whereas natural chromosome ends were not reactive in this way, illustrating how a double-strand break can behave differently from a telomere despite both being the ends of linear DNA molecules. Breakage–fusion–bridge cycles can lead to gross changes in the karyotype, and preventing them from occurring is one way that a telomere contributes to chromosome stability. Molecular studies using cloned telomeric sequences, for example in *Paramecium* (Bourgain and Katinka 1991), have confirmed that one role of telomeres is to prevent end fusions.

Because of the genome damage that results from the reactive nature of double-strand breaks many species have developed monitoring systems to arrest cell division in response to such breaks, allowing time for them to be repaired before proceeding into mitosis. For example, in the yeast *Saccharomyces cerevisiae* a single double-strand break *in vivo* is sufficient to cause cell cycle arrest (Brown *et al.* 1991). The yeast system which monitors the presence of double-strand breaks and causes cell cycle arrest requires the product of the *RAD9* gene, without which the cells show increased radiation sensitivity. By arresting the cell cycle until the break is repaired checkpoints such as *RAD9* prevent the large-scale damage to the genome that would occur if a double-strand break instigated breakage–fusion–bridge cycles. Similarly, human cells have a monitoring system for genomic damage involving the p53 gene product (see Chapter 7).

Classic studies on the fate of specific HO endonuclease-induced breaks at the yeast mating-type locus indicated that a non-essential chromosome terminating in a double-strand break was very unstable and rapidly lost (Weiffenbach and Haber 1981; Klar *et al.* 1984). Similar chromosome loss results from an HO break at a non-*MAT* locus (Sandell and Zakian 1993) One role of the *RAD9* checkpoint is illustrated by the fate of cells containing an HO-induced double-strand break (Bennett *et al.* 1993). Rad$^+$ and *rad9* strains containing an inducible *HO* gene and a dispensable plasmid carrying a target site for HO cleavage were studied. Induction of *HO* caused G$_2$ arrest in the Rad$^+$ strains, a cell cycle arrest phenotype similar to that seen for γ-irradiated Rad$^+$ cells. However, a different effect was observed with the *rad9* cells. Despite growth on medium which did not select for retention of the cut plasmid, only 30% of the cells plated yielded colonies following *HO* induction, and all of these were the result of plasmid loss. The remaining 70% produced microcolonies of dead cells arrested at various points in the cell cycle and showing a wide range of abnormal morphologies (enlarged vacuoles, altered nuclear morphology). Although there is no direct evidence, it seems that not only is a chromosome terminating in a double-strand break not maintained for very many divisions, but that it also produces a marked pleiotropy on the rest of the cell. Because the cut plasmid carried a centromere one possibility is that such a broken chromosome in a *rad9* strain fuses with other chromosomes and instigates breakage–fusion–bridge cycles, with the subsequent genome rearrangements leading to cell death. These experiments illustrate that if not repaired not only can a broken chromosome not be maintained but it can also cause damage to other cellular functions.

Organization in the mitotic interphase nucleus

Chromosomes occupy discrete domains in the mitotic interphase nucleus

There is perhaps a tendency for molecular biologists, familiar with dealing with DNA in solution, to view the interphase nucleus as a 'ball of string'. The rather undistinguished appearance of the mitotic interphase nucleus of many species viewed by light or electron microscopy might lead one to assume there is little if any order within it, and that the DNA strands of each chromosome are randomly distributed throughout the entire volume of the nucleus. However, there is considerable evidence suggesting that the interphase nucleus is far from lacking in large-scale order (reviewed by Hilliker and Appels 1989; Spector 1993). A number of biological processes are compartmentalized in the interphase nucleus. We are, for example, familiar with the concept of ribosomal DNA and ribosome biogenesis being localized to specific regions of the nucleus, the nucleoli (Heitz 1931; McClintock 1934; Kaplan *et al.* 1993). RNA splicing also appears to be compartmentalized to a small number of foci in the nucleus (Lamond and Carmo-Fonseca 1993).

The concept that chromosomes are not randomly distributed in the nucleus but maintain relatively fixed positions, remaining as discrete entities throughout interphase, goes back to the turn of the century. The reproducible position from one cell cycle to the next of visible features such as nuclear lobes in *Parascaris* eggs was suggested to be the result of chromosomes maintaining relatively fixed positions in the nucleus and remaining as discrete entities throughout interphase (Boveri 1888; reviewed by Wilson 1925; Comings 1980). In some specialized situations discrete chromosomes are visible in the interphase nuclei of mitotic cells, such as the polytene chromosomes of *Drosophila* salivary gland nuclei. In this example the centromeres are seen clustered at one end of the nucleus, with the telomeres adjacent to the nuclear envelope at the other (Agard and Sedat 1983; Mathog *et al.* 1984; Hochstrasser *et al.* 1986; Hiraoka *et al.* 1990). Such an orientation was first reported by Rabl (1885) who observed that recently condensed salamander prophase chromosomes are polarized, as if the position at the end of the previous telophase had been maintained throughout the intervening interphase despite decondensation of the chromosomes. The polarized telophase orientation of chromosomes, with the centromeres at one side and the telomeres at the other, reflects the force applied at the centromere by interaction with the mitotic spindle during anaphase chromosome movement. A Rabl orientation has been reported in many species (reviewed by Comings 1980) although it is by no means universal. For example, it is generally not found in mammalian cells (Manuelidis and Borden 1988; Billia and De Boni 1991).

The telophase chromosome organization is the starting point for the organization of the interphase nucleus. The Rabl organization in interphase salivary gland nuclei in *Drosophila* suggests that dramatic reorganization does not occur. Indeed, using time-lapse microscopy of living cells, Hiraoka *et al.* (1989) have been able to observe diploid nuclei as chromosomes decondense at the end of telophase, pass through interphase, and condense at the start of the next prophase. This revealed that chromosome organization is largely maintained in the intervening interphase.

One way to visualize chromosomes in interphase is by *in situ* hybridization, for example using chromosome-specific probes that decorate entire chromosomes. It is well established that chromosomes occupy distinct and relatively compact territories within the interphase nuclei of both plants and animals (Avivi and Feldman 1980; Schardin *et al.* 1985; Dyer *et al.* 1989; van Dekken *et al.* 1989; Heslop-Harrison and Bennett 1990; Haaf and Schmid 1991; Ferguson and Ward 1992). This organization is non-random and can depend on cell type (Borden and Manuelidis 1988). There is a large body of cytological evidence to argue that chromosomes exist as only slightly decondensed entities in the interphase nucleus. Together with the maintenance of chromosome domains this may facilitate chromosome condensation during prophase, which might be much more difficult if all the chromosomes were interwoven with each other in interphase.

Telomeres and the nuclear envelope in mitotic cells

There is a large body of indirect evidence to suggest that there are attachments between telomeres in interphase, and between telomeres and the nuclear envelope (Hilliker and Appels 1989). However, many of the data do not directly concern mitotic interphase nuclei but instead use polytene or meiotic chromosomes, or instances where telomere–telomere interactions are so strong that they are maintained in squash preparations of prophase nuclei (Wagenaar 1969; Ashley 1979; Avivi and Feldman 1980). The three dimensional organization of telomeres and indeed whole chromosomes can now be directly analysed by the use of *in situ* hybridization techniques coupled with fixation protocols which minimize disruption of the existing organisation and optical sectioning techniques such as laser confocal microscopy. This has been successfully applied in a number of species, including plants, mammals, trypanosomes and fission yeast. *In situ* hybridization to detect $(TTAGGG)_n$ has been used to determine the organization of mouse telomeres (Billia and De Boni 1991; Vourc'h *et al*. 1993), where they were found not to be associated with the nuclear envelope but distributed throughout the volume of the interphase nucleus. Similarly, human telomeres do not appear to be associated with the nuclear envelope but are distributed throughout the nuclear volume (Manuelidis and Borden 1988; Ferguson and Ward 1992). Walker *et al*. (1991) have observed that the inactive human X chromosome forms itself into a loop close to the nuclear envelope with the two telomeres very close together, in contrast to the linear arrangement of the active X chromosome, possibly reflecting a telomere–telomere association. Rawlins *et al*. (1991) have used *in situ* hybridization to detect the telomeric $(TTAGGG)_n$ sequences at the telomeres of the plants *Vicia faba* and *Pisum sativum*, where the signal was found predominantly next to the nuclear envelope, with the exception of a few adjacent to the nucleoli. Finally, in the majority of trypanosomes observed by Chung *et al*. (1990) the $(TTAGGG)_n$ signal was found distributed throughout the volume of the nucleus.

The organization of fission yeast telomeres in the interphase nucleus has been analysed directly by *in situ* hybridization (Funabiki *et al*. 1993) where they were found close to the nuclear periphery, although such analysis is hampered by the relatively small nucleus. *In situ* hybridization to detect telomeric sequences has not yet been successful in budding yeast, although there is circumstantial evidence for telomere location inferred from the location of the telomere-associated RAP1 protein (Chapter 5). This protein is a transcriptional regulator of a number of yeast genes and binds numerous sites in the yeast genome, including the terminal TG_{1-3} repeat arrays. Although metaphase chromosomes cannot be visualized in *S. cerevisiae*, meiotic chromosomes labelled with anti-RAP1 antibodies show a predominantly, although not exclusively, telomeric labelling pattern (Klein *et al*. 1992; Gilson *et al*. 1993). The RAP1 protein in interphase mitotic nuclei locates to around eight foci close to the nuclear periphery. As there are over 30 telomeres in a haploid cell this has been taken to imply that actual chromosome ends are located in clusters at the nuclear periphery, perhaps associated with the nuclear envelope. Various mutations alter the distribution of the RAP1 protein in the interphase nucleus, and from this it has been argued that these mutations cause a similar redistribution of chromosome ends (Palladino *et al*. 1993; see Chapter 8). However, it should be emphasized that there is no direct evidence that the observed RAP1 distribution corresponds to the location of telomeric DNA sequences in yeast. The only direct evidence that yeast telomeres are attached to the nuclear envelope comes from electron microscopy of serial thin sections of meiotic nuclei, which do show such attachments (Byers and Goetsch 1975).

In *Drosophila* the telomeres are clustered together at one end of the salivary gland nucleus and attached to the nuclear envelope (see above).

Does this reflect a specific interaction between telomeres and the nuclear envelope or simply that the organization of chromosomes at telophase is not substantially changed during interphase? Although this is a difficult question to address, there is one potential mechanism that might stabilize the telophase organization in interphase. Interphase chromatin is anchored by many sites to the nuclear matrix, which may prevent large-scale interphase chromosome movement (Jackson 1991). Furthermore, the nuclear envelope is lined by lamins; these do not appear to have a preference for telomeric regions and for example can bind all along condensed chromosomes *in vitro* (Glass and Gerace 1990; reviewed by Moir and Goldman 1993). When the nuclear envelope reforms around the group of chromosomes the lamina is likely to form interactions with those chromosomal regions at the periphery of the chromosome group. Because of polarization, the telomeres are likely to be located at the edge of the chromosome group at telophase and thus may preferentially form an interaction with the lamina as the nucleus reforms. If stable this interaction may be sufficient to maintain the telomeres in a peripheral location throughout interphase, giving the appearance of a specific telomere–nuclear envelope interaction.

Support for this speculation comes from diploid chromosome behaviour in *Drosophila* (Hiraoka *et al.* 1990). In early prophase, before nuclear envelope breakdown, the chromosomes are arranged in a Rabl orientation with the telomeres clustered together. However, as soon as the cells progress into early metaphase and nuclear envelope breakdown starts, the telomere clustering is lost. These telomere–telomere associations can also be seen as thin strands of chromatin connecting non-homologous chromosomes in squash preparations (e.g. Berendes and Meyer 1968; Rubin 1978; Young *et al.* 1983). Similar telomere–telomere associations have been seen in prophase squash preparations in plants (Wagenaar 1969; Ashley 1979; Avivi and Feldman, 1980). Again they are seen only in prophase, before nuclear envelope breakdown, and are lost by the time the cells have progressed to metaphase. The loss of these associations may suggest that they do not reflect a direct interaction between telomeres, but instead one that is mediated by the proximity of their attachments to the nuclear envelope, and that it is the nuclear envelope and lamina which are holding the chromosome ends together. In support of this speculation is the observation that disrupting the yeast nuclear envelope with detergent is sufficient to increase the number of RAP1 foci from around 8–16 to approximately 60 in a diploid cell, as if clustering of the telomeres has been affected (Klein *et al.* 1992; Gilson *et al.* 1993).

Another possibility is that telomere–telomere interactions reflect the unusual properties of the constitutive heterochromatin that is found at the telomeres of numerous plant and animal species (see Chapter 3). In general, organisms which show blocks of telomeric constitutive heterochromatin show telomeric associations of that heterochromatin (Wagenaar 1969;

Fussell 1977; Barnes *et al.* 1985). In contrast, species which lack such terminal heterochromatin (e.g. most human chromosomes) do not generally show such associations. For example, the large number of (TTAGGG)$_n$ signals in the mammalian interphase nucleus (Billia and De Boni 1991) indicates that mammalian telomeres are not clustered together into a small number of foci. Electron microscopy has established that in many eukaryotes constitutive heterochromatic regions of the interphase genome, both terminal and interstitial, are often located close to the nuclear periphery. Thus if a species has terminal blocks of heterochromatin this may lead to location of the telomeres at the nuclear periphery. It is, therefore, difficult to determine if the observed associations between telomeres and with the nuclear envelope in some species reflects the behaviour of telomeres *per se* or constitutive heterochromatin.

It remains unclear if there is a specific interaction between the nuclear envelope and terminal regions of chromosomes and the molecular details are not known. The molecular nature of telomere–telomere interactions is also not known, although there are some potential model systems. Some ciliate telomere binding proteins aggregate *in vitro*, which may have relevance to telomere–telomere associations in these species (discussed in Chapter 5). Telomere–telomere interactions have also been suggested to be the result of DNA–DNA interactions involving secondary structures stabilized by G-tetrads (Chapter 3). However, the *in vivo* telomere associations in these species appear to be protease-sensitive, suggesting instead that they are mediated by protein–protein interactions. In yeast there is evidence for a transient association between the DNA at chromosome ends during S phase, the molecular nature of which remains unknown (Wellinger *et al.* 1992, 1993*a,b*; see Chapter 3).

Implications of mitotic interphase genome organization for gene expression and replication

Telomere–telomere interactions, together with attachments to the nuclear envelope, might contribute to chromosome organization in the nucleus. This in turn might have implications for processes such as gene expression and DNA replication (Jackson 1990, 1991; Manuelidis 1990). The organization of whole chromosome domains in the nucleus is not totally random, although it may in part reflect mechanical considerations of segregation and chromosome condensation. Are large-scale changes in chromosome position used as a mechanism of regulating gene expression *in vivo*? Although few experiments to address this question directly have been performed there is a certain amount of circumstantial evidence. For example, most balanced chromosome translocations have no phenotype, even though they presumably disrupt the interphase domains of the chromosomes involved. Furthermore, chromosomes are continually rearranging through evolution;

in wild mouse populations Robertsonian fusions are common, yet have little obvious effect on gene expression. The normal function of genes on single human chromosomes in human–mouse somatic cell hybrids also argues against a role for changes in large-scale chromosome order as a widespread mechanism for the regulation of gene expression.

In *Drosophila* chromosome organization is not completely random and is characteristic of the tissue being studied (Hochstrasser and Sedat 1987*a*). However, no euchromatic locus has an invariant location in the salivary gland nucleus (Mathog *et al*. 1984). Hochstrasser and Sedat (1987*b*) have analysed chromosome positioning in response to changes in gene expression caused by heat shock or ecdysone treatment. No changes in chromosome positions were found, arguing that chromosomal loci are not repositioned in response to, or prefigure, changes in transcription. Instead, the configurations of a number of rearranged chromosomes has led Mathog and Sedat (1989) to argue that chromosome organization within the salivary gland nucleus arises largely from the mechanics of mitosis, somatic pairing, and endoreduplication. There is little evidence that changes in chromosome position is used as a mechanism for differentially regulating euchromatic genes in *Drosophila*.

Telomere behaviour in meiosis

Three key features of meiosis are a homology search leading to pairing of homologues, formation of a ribbon-like structure called the synaptonemal complex (SC), and reciprocal exchanges which, when done in the context of the SC, are seen as chiasmata (Westergaard and von Wettstein 1972; von Wettstein *et al*. 1984). Until recently the relationship between pairing, SC formation, and recombination has been inferred from cytological observations. However, using yeast genetics to manipulate these processes experimentally has led to a number of unexpected results, and the relationship between these various processes is currently being reassessed (reviewed by Hawley and Arbel 1993).

Although the cytological aspects of pairing in meiotic prophase have been extensively studied, the underlying molecular mechanisms are poorly understood. In most species with short meiotic chromosomes, pairing and SC formation usually initiates at sites close to, but not precisely at, the ends of the chromosomes (Gillies 1975; Rasmussen and Holm 1980; Loidl 1990). As meiosis continues the SC extends progressively further along the chromosomes until finally the entire length of the chromosome is paired. In species with much longer meiotic chromosomes, such as the enormous chromosomes seen in plants such as *Allium, Tradescantia*, and lilies, there can also be a large number of interstitial initiation sites.

The three-dimensional organization of meiotic chromosomes can be reconstructed from electron micrographs of serial thin sections through meiotic nuclei. Applying this elegant if laborious technique to a number of species has shown that the telomeres are attached to the nuclear envelope, often by prominent densely staining material termed attachment plaques (Gillies 1975; Rasmussen and Holm 1980). In most species studied both telomeres of each chromosome are attached, although in nematodes such as *Caenorhabditis elegans* and *Ascaris* only one chromosome end is attached. Although less common than attachment to the nuclear envelope, in some species these attachment sites become clustered in one region of the nuclear envelope, causing the chromosomes to adopt a configuration in the nucleus termed the bouquet formation; it is not seen in a number of species including *Coprinus* and *Neurospora* (von Wettstein *et al*. 1984; Loidl 1990). It has been suggested that one consequence of this is that chromosomes are helped to align (Moses 1968; Gilies 1975; Rasmussen and Holm 1980). Because of their attachment to the nuclear envelope, sequences near the telomeres are restricted to a search through a two-dimensional plane for their homologous telomere with which to pair, whereas internal chromosomal regions are free to explore three-dimensional space. The clustering of telomeres which produces the bouquet formation may further facilitate this process. The preferential initiation of pairing near telomeres may therefore occur because such pairings are favoured kinetically. By helping the telomeres to align the pairing of the whole chromosome is facilitated, as SC formation can continue progressively along the chromosome. The initial telomeric alignments of mammalian chromosomes are often incorrect (Chandley, 1989) which would be consistent with the high level of sequence variability at mammalian telomeres (Chapter 10).

Attachment of meiotic telomeres to the nuclear envelope is an almost universal feature and in theory should facilitate chromosome alignment and pairing in telomeric regions (reviewed by Loidl 1990). However, telomeric initiation is not an absolute requirement for pairing and SC formation. Even in species where telomeric initiation of pairing is the norm, interstitial initiation can occur (Gillies 1975). Ring chromosomes, which do not have telomeres, are also capable of orderly synapsis. It is interesting to also consider cases of inversion chromosomes and in particular heterozygotes where a segment of one homologue has been inverted (e.g. de Perdigo *et al*. 1989). If telomeric initiation and subsequent processive formation of the SC in a linear fashion along the chromosome is the sole determinant of pairing then the lack of homology in the inverted region should lead to either no pairing (asynapsis) or linear pairing of the antiparallel segments despite the lack of sequence homology (heterosynapsis). However, although it may accelerate the process, telomeric alignment is not the sole criterion for synapsis because inversion loops can sometimes be formed in such heterozygotes in a number of species such as the mouse (Moses *et al*. 1982) and man (de Perdigo *et al*. 1989). In such examples an internal initiation has occurred in the inverted region and the chromosomes are in a looped formation, with all the chromosomal regions aligned on the basis of homology rather than linearity (homosynapsis). Although homologous synapsis in the inverted region results in a loop formation in zygotene and early pachytene, such inversion loops are often lost at the later stages of meiosis by what is termed synaptic adjustment (Moses *et al*. 1982).

Distribution of chiasmata and implications for genetic maps

Human meiotic chromosomes are longer in females, which corresponds to a similar increase in the apparent genetic length in female meiosis as determined by linkage analysis. When male and female genetic maps are compared a map expansion in the terminal regions is seen, although this is not a dramatic effect. Despite the overall map length being greater in females, the recombination frequency between markers in terminal regions is higher in males than in females (see Chapter 10). The map expansion in terminal regions of males presumably reflects subtle differences in the processes of pairing and recombination during male and female meiosis.

Cytologically visible chiasmata have long been understood to be the direct result of physical exchange between homologues. Scoring chiasma position and frequency has therefore been regarded as a straightforward way of measuring crossovers in the genome (Nilsson *et al*. 1993). Such an approach is not possible in well studied genetic organisms such as *C. elegans*, yeast, or *Drosophila* because they do not have meiotic chromosomes suitable for such a study. Similarly, most species for which there are good cytological data on the frequency and distribution of chiasmata do not have a high quality genetic map anchored onto a physical map. Although genetic maps are now becoming available for a number of plant species which can be compared with the known cytological data (Nilsson *et al*. 1993), extensive cytological data in combination with good genetic data are really only available for the human male. The distribution of chiasmata has been extensively studied and can now be correlated with the emerging physical and genetic maps (see Chapter 10). One question that arises is that of the apparent high frequency of terminal chiasmata.

Cytological studies (Hultén 1974) revealed that for most human chromosomes there is a high frequency of chiasmata in the most terminal position. For example, for chromosome 16q in spermatocytes around 90% of the chiasmata scored were in the most terminal 10% of the length of the arm. Such end-to-end associations are a feature of a number of species. One possibility is that this reflects chiasma terminalization, that is the formation of a chiasma at an internal location which then moves along the chromosome to the end. The concept of terminalization of chiasmata was first advanced by Darlington (1929). However, this hypothesis has been criticized; there is now extensive evidence that chiasma position generally correlates with a crossover and that neither terminalization nor any significant movement of chiasmata occurs in most species (reviewed by John 1990). This can be illustrated by differentially labelling sister chromatids by BrdU incorporation and Giemsa staining. Internal chiasmata clearly correspond to crossover events in the mouse (Kanda and Kato 1980) and *Allium* (Loidl 1979). For bivalents with only terminal 'chiasmata' an internal strand exchange should be visible if terminalization has occurred, yet this has not been observed. Although such 'chiasmata' do not reflect terminalization, there is no stainable material beyond the end of the chromosome and it is therefore not possible to determine if such terminal 'chiasmata' are associated with a crossover event by cytological criteria. It is therefore possible that such apparent terminal 'chiasmata' do not in fact reflect sites of genetic crossover and are achiasmate terminal adhesions between homologous chromosomes.

There are preliminary data which make it possible to test this hypothesis by comparing the distributions of chiasmata and genetic linkage maps. For human chromosome 21q the two maps correspond well for most of their length (Hultén *et al.* 1990). The most terminal 10% of 21q shows a 4-fold increased chiasma frequency as compared with other regions of the chromosome, and there is a corresponding terminal map expansion of the linkage map. Thus for chromosome 21q the distribution of chiasmata may provide a reasonable if low-resolution indication of the genetic linkage map. However, this is a chromosome which does not show the very dramatic terminal chiasma frequency seen for other chromosomes. For most chromosomes if the chiasma distribution reflected recombination events one would predict that the terminal 10% of the physical length of each chromosome would correspond to around 90% of the genetic length of the chromosome (Hultén, 1974). The human genetic map has yet to reveal such a large expansion in the terminal regions of the chromosomes, even though the genetic maps of a number of chromosomes extend to the telomere (NIH/CEPH Collaborative Mapping Group, 1992). For example, the genetic map of the human X chromosome in female meiosis does not reveal a particularly high rate of recombination in the terminal region (Henke *et al.* 1993). Genetic markers have been obtained which extend to within 30 kb of the Xp telomere, yet the terminal 2 Mb of the chromosome has a female genetic length of only 5.2 cM, which is not dissimilar to the per megabase figure for the X chromosome as a whole. Another example of where the linkage map extends to the telomere is chromosome 7q (Helms *et al.* 1992) and again there is a modest terminal map expansion but not the dramatic increase in recombination rates that would be predicted from the chiasma data. Similarly, the genetic map of the mouse does not reveal the massive map expansion in the terminal regions which cytological studies might suggest (Ashley *et al.* 1993), even though a number of genetic markers very close to chromosomes ends have been obtained (see Chapter 10). As the mouse and human genetic maps improve over the next few years a more detailed picture of recombination within terminal regions will emerge. The current data do suggest that the chiasma data for the terminal regions may not accurately reflect recombination events, and we must therefore consider the possibility that these apparent terminal 'chiasmata' do not reflect crossover events and might instead reflect a distinct telomere–telomere association.

Why might there be a second pathway to promote association between homologous chromosomes? Chiasmata have an essential segregational role in mammalian meiosis. A bivalent is a pair of homologous chromosomes joined by one or more chiasmata. The latter are essential to balance the spindle forces on the kinetochores of the homologous chromosomes, thus ensuring a stable bipolar spindle attachment at metaphase I. In contrast, the monopolar attachment of univalents is unstable and they will not segregate correctly. In some species such as *Drosophila* there is an additional system (distributive disjunction) to ensure segregation of achiasmate chromosomes (reviewed by Hawley and Theurkauf 1993). This process appears to be rather crude, with chromosomes segregated away from each other on the basis of size and shape. Such a system may be too inaccurate for species such as the mouse with a large number of similar sized and shaped chromosomes.

Summary

Attachment of telomeres with the nuclear envelope is a widespread phenomenon in meiosis. In theory it may facilitate the process of chromosome pairing by providing a very rapid alignment of the chromosome ends. However, the behaviour of ring chromosomes and the capability of inversion heterozygotes to form inversion loops argues that telomere attachment to the nuclear envelope and telomeric initiation cannot be an indispensable requirement for chromosome pairing. Nevertheless, its widespread occurrence, in contrast to mitotic telomere attachment to the nuclear envelope, argues for a role for these attachments in meiosis, even if only supporting. Despite their potential importance, the molecular details of telomeric attachments to the nuclear envelope remain unknown.

It seems essential that all chromosomes have at least one chiasma if high rates of formation of aneuploid gametes are to be avoided, as the lack of a chiasma will lead to nondisjunction at metaphase I. The high frequency of terminal 'chiasmata' in humans may not necessarily reflect genetic crossover events. A number of authors have considered the possibility that these terminal associations which appear to be 'chiasmata' do not reflect crossover events (John 1990). One possibility is that they are a specialized terminal association which holds homologous chromosomes together even without a crossover event and thus allows normal segregation at metaphase I. This speculation will be tested as the genetic maps of both the human and mouse improve over the coming years, but we should consider the possibility that telomeres have a specialized role in meiotic chromosome segregation by causing end-to-end associations which currently are classified as chiasmata.

The cytological observations that terminal deletions and inversions were not readily recovered in *Drosophila* first defined the telomere. A chromosome without this specialized structure was observed to be highly reactive, leading to damage to the genome following breakage–fusion–bridge cycles. A telomere is whatever biological structure makes the end of a natural chromosome behave differently from a double-strand break. Such a functional definition serves as the basis for detailed molecular studies. By defining a telomere as a structure with biological functions we can then proceed to investigate what DNA sequences and interacting proteins contribute to this biological behaviour.

References

Reviews

Avivi, L. and Feldman, M. (1980). Arrangement of chromosomes in the interphase nucleus of plants. *Human Genet.*, **55**, 281–95.

Comings, D. E. (1980). Arrangement of chromatin in the nucleus. *Human Genet.*, **53**, 131–143.

Gillies, C. B. (1975). Synaptonemal complex and chromosome structure. *Annu. Rev. Genet.*, **8**, 91–109.

Gilson, E., Laroche, T., and Gasser, S. M. (1993). Telomeres and the functional architecture of the nucleus. *Trends Cell Biol.*, **3**, 128–34.

Haaf, T. and Schmid, M. (1991). Chromosome topology in mammalian interphase nuclei. *Exp. Cell Res.*, **192**, 325–32.

Hawley, R. S. and Arbel, T. (1993). Yeast genetics and the fall of the classical view of meiosis. *Cell*, **72**, 301–3.

Hawley, R. S. and Theurkauf, W. E. (1993). Requiem for distributive segregation: achiasmate segregation in *Drosophila* females. *Trends Genet.*, **9**, 310–17.

Heslop-Harrison, J. S. and Bennett, M. D. (1990). Nuclear architecture in plants. *Trends Genet.*, **6**, 401–5.

Hilliker, A. J. and Appels, R. (1989). The arrangement of interphase chromosomes: structural and functional aspects. *Exp. Cell Res.*, **185**, 297–318.

Jackson, D. A. (1990). The organisation of replication centres in higher eukaryotes. *Bioessays*, **12**, 87–9.

Jackson, D. A. (1991). Structure–function relationships in eukaryotic nuclei. *Bioessays*, **13**, 1–10.

John, B. (1990). *Meiosis*. Cambridge University Press, Cambridge.

Lamond, A. I. and Carmo-Fonseca, M. (1993). The coiled body. *Trends Cell Biol.*, **3**, 198–204.

Loidl, J. (1990). The initiation of meiotic chromosome pairing: the cytological view. *Genome*, **33**, 759–78.

Manuelidis, L. (1990). A view of interphase chromosomes. *Science*, **250**, 1533–40.

Moir, R. D. and Goldman, R. D. (1993). Lamin dynamics. *Curr. Opin. Cell Biol.*, **5**, 408–11.

Moses, M. J. (1968). Synaptinemal complex. *Annu. Rev. Genet.*, **2**, 363–412.

Nilsson, N.-O., Säll, T. and Bengtsson, B. O. (1993). Chiasma and recombination data in plants: are they compatible? *Trends Genet.*, **9**, 344–8.

Rasmussen, S. W. and Holm, P. B. (1980). Mechanics of meiosis. *Hereditas*, **93**, 187–216.

Spector, D. L. (1993). Macromolecular domains within the cell nucleus. *Annu. Rev. Cell Biol.*, **9**, 265–315.

von Wettstein, D., Rasmussen, S. W., and Holm, P. B. (1984). The synaptonemal complex in genetic segregation. *Annu. Rev. Genet.*, **18**, 331–413.

Westergaard, M. and von Wettstein, D. (1972). The synaptinemal complex. *Annu. Rev. Genet.*, **6**, 71–110.

Wilson, E. B. (1925). *The cell in development and heredity*. Macmillan, New York.

Primary papers

Agard, D. A. and Sedat, J. W. (1983). Three-dimensional architecture of a polytene nucleus. *Nature*, **302**, 676–81.

Ashley, T. (1979). *Specific* end-to-end attachment of chromosomes in *Ornithogalum virens*. *J. Cell Sci.*, **38**, 357–67.

Ashley, T., Cacheiro, N. L. A., Russell, L. B., and Ward, D. C. (1993). Molecular characterization of a pericentric inversion in mouse chromosome 8 implicates telomeres as promoters of meiotic recombination. *Chromosoma*, **102**, 112–120.

Barnes, S. R., James, A. M., and Jamieson, G. (1985). The organization, nucleotide sequence, and chromosomal distribution of a satellite DNA from *Allium cepa*. *Chromosoma*, **92**, 185–92.

Bennett, C. B., Lewis, A. L., Baldwin, K. K., and Resnick, M. A. (1993). Lethality induced by a single site-specific double-strand break in a dispensable yeast plasmid. *Proc. Natl. Acad. Sci. USA*, **90**, 5613–17.

Berendes, H. D. and Meyer, G. F. (1968). A specific chromosome element, the telomere of *Drosophila* polytene chromosomes. *Chromosoma*, **25**, 184–97.

Billia, F. and De Boni, U. (1991). Localization of centromeric satellite and telomeric DNA sequences in dorsal root ganglion neurons, *in vitro*. *J. Cell Sci.*, **100**, 219–26.

Borden, J. and Manuelidis, L. (1988). Movement of the X chromosome in epilepsy. *Science*, **242**, 1687–91.

Bourgain, F. M. and Katinka, M. D. (1991). Telomeres inhibit end to end fusion and enhance maintenance of linear DNA molecules injected into the *Paramecium primaurelia* macronucleus. *Nucleic Acids Res.*, **19**, 1541–7.

Boveri, T. (1888) Die Befruchtung und Teilung des Eies von *Ascaris megalocephala*. In: (ed. Zellen-Studien, H.) G. Fischer, Jena, Vol. 2, pp. S. 1–189.

Brown, M., Garvik, B., Hartwell, L., Kadyk, L., Seeley, T., and Weinert, T. (1991). Fidelity of mitotic chromosome transmission. *Cold Spring Harbor Symp. Quant. Biol.*, **56**, 359–65.

Byers, B. and Goetsch, L. (1975). Electron microscopic observations on the meiotic karyotype of diploid and tetraploid *Saccharomyces cerevisiae*. *Proc. Natl. Acad. Sci. USA*, **72**, 5056–60.

Chandley, A. C. (1989). Asymmetry in chromosome pairing: a major factor in de novo mutation and the production of genetic disease in man. *J. Med. Genet.*, **26**, 546–52.

Chung, H.-M. M., Shea, C., Fields, S., Taub, R. N., Van der Ploeg, L. H. T., and Tse, D. B. (1990). Architectural organization in the interphase nucleus of the protozoan *Trypanosoma brucei*: location of telomeres and mini-chromosomes. *EMBO J.*, **9**, 2611–19.

Darlington, C. D. (1929). Chromosome behavior and structural hybridity in *Tradescantia. J. Genet.*, **21**, 207–86.

de Perdigo, A., Gabriel-Robez, O., and Rumpler, Y. (1989). Correlation between chromosomal breakpoint positions and synaptic behaviour in human males heterozygous for a pericentric inversion. *Human Genet.*, **83**, 274–6.

Dyer, K. A., Canfield, T. K., and Gartler, S. M. (1989). Molecular cytological differentiation of active from inactive X domains in interphase: implications for X chromosome inactivation. *Cytogenet. Cell Genet.*, **50**, 116–20.

Ferguson, M. and Ward, D. C. (1992). Cell cycle dependent chromosomal movement in pre-mitotic human T-lymphocyte nuclei. *Chromosoma*, **101**, 557–65.

Funabiki, H., Hagan, I., Uzawa, S., and Yanagida, M. (1993). Cell cycle-dependent specific positioning and clustering of centromere and telomeres in fission yeast. *J. Cell. Biol.*, **121**, 961–76.

Fussell, C. P. (1977). Telomere associations in interphase nuclei of *Allium cepa* demonstrated by C-banding. *Exp. Cell Res.*, **110**, 111–17.

Glass, J. R. and Gerace, L. (1990). Lamins A and C bind and assemble at the surface of mitotic chromosomes. *J. Cell Biol.*, **111**, 1047–57.

Heitz, E. (1931). Die ursache der gesetzmassigen zahl, lage, form und grosse planzlicher nukleolen. *Planta*, **12**, 775–84.

Helms, C., Mishra, S. K., Riethman, H., Burgess, A. K., Ramachandra, S., Tierney, C. et al. (1992). Closure of a genetic linkage map of human chromosome 7q with centromere and telomere polymorphisms. *Genomics*, **14**, 1041–54.

Henke, A., Fischer, C., and Rappold, G. A. (1993). Genetic map of the human pseudoautosomal region reveals a high rate of recombination in female meiosis at the Xp telomere. *Genomics*, **18**, 478–85.

Hiraoka, Y., Minden, J. S., Swedlow, J. R., Sedat, J. W., and Agard, D. A. (1989). Focal points for chromosome condensation and decondensation revealed by three-dimensional *in vivo* time-lapse microscopy. *Nature*, **342**, 293–6.

Hiraoka, Y., Agard, D. A., and Sedat, J. W. (1990). Temporal and spatial coordination of chromosome movement, spindle formation, and nuclear envelope breakdown during prometaphase in *Drosophila melanogaster* embryos. *J. Cell Biol.*, **111**, 2815–28.

Hochstrasser, M. and Sedat, J. W. (1987a). Three-dimensional organization of *Drosophila melanogaster* interphase nuclei. I. Tissue-specific aspects of polytene nuclear architecture. *J. Cell Biol.*, **104**, 1455–70.

Hochstrasser, M. and Sedat, J. W. (1987b). Three-dimensional organization of *Drosophila melanogaster* interphase nuclei. II. Chromosome spatial organization and gene regulation. *J. Cell Biol.*, **104**, 1471–83.

Hochstrasser, M., Mathog, D., Gruenbaum, Y., Saumweber, H., and Sedat, J. W. (1986). Spatial organisation of chromosomes in the salivary gland nuclei of *Drosophila melanogaster. J. Cell Biol.*, **102**, 112–23.

Hultén, M. A. (1974). Chiasma distribution at diakinesis in the normal human male. *Hereditas*, **76**, 55–78.

Hultén, M., Lawrie, N. M., and Laurie, D. A. (1990). Chaisma-based genetic maps of chromosome 21. *Am. J. Med. Genet.* (Suppl.), **7**, 148–54.

Kanda, N. and Kato, H. (1980). Analysis of crossing over in mouse meiotic cells by BrdU labelling technique. *Chromosoma*, **78**, 113–21.

Kaplan, F. S., Murray, J., Sylvester, J. E., Gonzalez, I. L., O'Connor, J. P., Doering, J. L., Muenke, M., Emanuel, B. S., and Zasloff, M. A. (1993). The topographic organization of repetitive DNA in the human nucleolus. *Genomics*, **15**, 123–32.

Klar, A. J. S., Strathern, J. N., and Abraham, J. A. (1984). Involvement of double-strand chromosomal breaks for mating-type switching in *Saccharomyces cerevisiae*. *Cold Spring Harbor Symp. Quant. Biol.*, **49**, 77–88.

Klein, F., Laroche, T., Cardenas, M. E., Hofmann, J. F.-X., Schweizer, D., and Gasser, S. M. (1992). Localization of RAP1 and topoisomerase II in nuclei and meiotic chromosomes of yeast. *J. Cell Biol.*, **117**, 935–948.

Loidl, J. (1979). C-band proximity of chiasmata and absence of terminalisation in *Allium flavum* (Liliaceae). *Chromosoma*, **73**, 45–51.

McClintock, B. (1934). The relation of a particular chromosomal element to the development of the nucleoli in Zea mays. *Z. Zellforsch.*, **21**, 294–328.

McClintock, B. (1938). The fusion of broken ends of sister half-chromatids following chromatid breakage at meiotic anaphase. *Res. Bull. -Mo., Agric. Exp. Stn.*, **290**, 1–48.

McClintock, B. (1939). The behavior of successive nuclear divisions of a chromosome broken at meiosis. *Proc. Natl. Scad. Sci. USA*, **25**, 405–416.

McClintock, B. (1941). The stability of broken ends of chromosomes in Zea mays. *Genetics*, **26**, 234–82.

Manuelidis, L. and Borden, J. (1988). Reproducible compartmentalization of individual chromosome domains in human CNS cells revealed by in situ hybridization and three-dimensional reconstruction. *Chromosoma*, **96**, 397–410.

Mathog, D. and Sedat, J. W. (1989). The three-dimensional organization of polytene nuclei in male *Drosophila melanogaster* with compound XY or ring X chromosomes. *Genetics*, **121**, 293–311.

Mathog, D., Hochstrasser, M., Gruenbaum, Y., Saumweber, H., and Sedat, J. (1984). Characteristic folding pattern of polytene chromosomes in *Drosophila* salivary gland nuclei. *Nature*, **308**, 414–21.

Moses, M. J., Poorman, P. A., Roderick, T. H., and Davisson, M. T. (1982). Synaptonemal complex analysis of mouse chromosomal rearrangements. *Chromosoma*, **84**, 457–74.

Muller, H. J., (1938). The remaking of chromosomes. *Collect. Net*, **8**, 182–195.

Muller, H. J. (1940). An analysis of the process of structural change in the chromosomes of *Drosophila*. *J. Genet.*, **40**, 1–66.

Muller, H. J. and Herskowitz, I. H. (1954). Concerning the healing of chromosome ends produced by breakage in *Drosophila meloanogaster*. *Am. Nat.*, **88**, 177–208.

NIH/CEPH Collaborative Mapping Group (1992). A comprehensive genetic linkage map of the human genome. *Science*, **258**, 67–86.

Palladino, F., Laroche, T., Gilson, E., Axelrod, A., Pillus, L., and Gasser, S. M. (1993). SIR3 and SIR4 proteins are required for the positioning and integrity of yeast telomeres. *Cell*, **75**, 543–555.

Rabl, C. (1885). Über Zellteilung. *Morphologisches Jahrbuch*, **10**, 214–330.

Rawlins, D. J., Highett, M. I., and Shaw, P. J. (1991). Localization of telomeres in plant interphase nuclei by in situ hybridization and 3D confocal microscopy. *Chromosoma*, **100**, 424–31.

Roberts, P. A. (1974). A cytogenetic analysis of X-ray induced 'visible' mutations at the *yellow* locus of *Drosophila melanogaster*. *Mutat. Res.*, **22**, 139–144.

Roberts, P. A. (1975). In support of the telomere concept. *Genetics*, **80**, 135–142.

Rubin, G. M. (1978). Isolation of a telomeric DNA sequence from *Drosophila melanogaster*. *Cold Spring Harbor Symp. Quant. Biol.*, **42**, 1041–6.

Sandell, L. L. and Zakian, V. A. (1993). Loss of a yeast telomere: arrest, recovery, and chromosome loss. *Cell*, **75**, 729–39.

Schardin, M., Cremer, T., Hager, H. D., and Lang, M. (1985). Specific staining of human chromosomes in Chinese hamster × man hybrid cell lines demonstrates interphase chromosome territories. *Human Genet.*, **71**, 281–87.

van Dekken, H., Pinkel, D., Mullikin, J., Trask, B., van den Engh, G., and Gray, J. (1989). Three-dimensional analysis of the organization of human chromosome domains in human and human–hamster interphase nuclei. *J. Cell Sci.*, **94**, 299–306.

Vourc'h, C., Taruscio, D., Boyle, A. L., and Ward, D. C. (1993). Cell cycle-dependent distribution of telomeres, centromeres, and chromosome-specific subsatellite domains in the interphase nucleus of mouse lymphocytes. *Exp. Cell Res.*, **205**, 142–51.

Wagenaar, E. B. (1969). End-to-end chromosome attachments in mitotic interphase and their possible significance to meiotic chromosome pairing. *Chromosoma*, **26**, 410–26.

Walker, C. L., Cargile, C. B., Floy, K. M., Delannoy, M., and Migeon, B. R. (1991). The Barr body is a looped X chromosome formed by telomere association. *Proc. Natl. Acad. Sci. USA.*, **88**, 6191–5.

Weiffenbach, B. and Haber, J. E. (1981). Homothallic mating type switching generates lethal chromosome breaks in *rad52* strains of *Saccharomyces cerevisiae*. *Mol. Cell. Biol.*, **1**, 522–34.

Wellinger, R. J., Wolf, A. J., and Zakian, V. A. (1992). Use of non-denaturing Southern hybridization and two dimensional agarose gels to detect putative intermediates in telomere replication in *Saccharomyces cerevisiae*. *Chromosoma*, **102**, S150–S156.

Wellinger, R. J., Wolf, A. J., and Zakian, V. A. (1993a). Saccharomyces telomeres acquire single-strand TG_{1-3} tails late in S phase. *Cell*, **72**, 51–60.

Wellinger, R. J., Wolf, A. J., and Zakian, V. A. (1993b). Origin activation and formation of single-strand TG_{1-3} tails occur sequentially in late S phase on a yeast linear plasmid. *Mol. Cell. Biol.*, **13**, 4057–65.

Young, B. S., Pession, A., Traverse, K. L., French, C., and Pardue, M. L. (1983). Telomere regions in Drosophila share complex DNA sequences with pericentric heterochromatin. *Cell*, **34**, 85–94.

3

Telomere structure

This chapter summarizes the nuclear telomere sequence organization for a number of eukaryotic species. In most cases tandem repeat arrays of a short (typically fewer than 10 nucleotides) sequence are found at the very end of the chromosome. These show considerable similarity in sequence between species and various lines of evidence implicate them in essential telomere function. In addition a wide range of repetitive sequences has been found in the subterminal regions of many species. These sequences are often species-specific and it has proved difficult to assign any biological role to them. Finally the ability of a number of telomeric sequences to fold into unusual secondary structures *in vitro*, stabilized by G·G bonds, will be discussed. However, before dealing with any individual species, methods to isolate and study telomeric sequences, and general conclusions regarding their structure, will be reviewed.

Isolation and characterization of telomeric sequences

Direct sequencing of genomic DNA

Much of the initial work on telomere structure centred on ciliates. These unicellular eukaryotes can have a vast number of telomeres per cell (over 10^7 for *Oxytricha nova*) as a result of programmed genome fragmentation and telomere addition (see Chapter 6). In contrast, a diploid human cell has a mere 92 telomeres and a much larger total genome size. In addition, the ends of ciliate chromosomes are so homogeneous in sequence that genomic DNA can be end-labelled and used directly for chemical sequencing. Unfortunately telomeres are too small a fraction of the genome of most species to be amenable to such an approach. However, if there is *a priori* evidence of what the terminal repeat sequence is likely to be, it is possible to obtain a small amount of sequence information by using a terminal repeat sequence oligonucleotide as a primer in a dideoxy sequencing reaction with genomic DNA as the template. This approach has been successfully used to obtain sequence information for tomato telomeres by using genomic DNA, size enriched for telomeres, and 5'-GGTTTAGGGTTTAG-3' as a sequencing primer (Ganal *et al.* 1991), based on the knowledge that other plants have $(TTTAGGG)_n$ as their terminal repeats.

Cross-hybridization as an approach to identifying telomeric sequences

Many terminal repeats are similar in sequence, and the repeats of one species will often cross-hybridize with those of another, especially under reduced stringency conditions. Such a repeat-containing probe might detect the ends of metaphase chromosomes by *in situ* hybridization or detect a BAL31-sensitive signal on Southern blots. However, although such hybridization data are useful to guide further work, they alone are not definitive proof of the sequence of the terminal repeats. For example, it is believed that all vertebrate telomeres are composed of $(TTAGGG)_n$ arrays, but the data for almost all species are based solely on *in situ* hybridization (Meyne *et al.* 1990). To illustrate the potential problem, a subtelomeric sequence has been identified in whalebone whales, of which there are 10 extant species including the blue, right, and minke whales. All have a G+C-rich satellite in their genome, a repetitive sequence consisting of large tandem repeat arrays of an underlying monomer unit that varies from around 400 to 1200 bp in length according to species. The monomers of all species consist of a unique sequence of 211 or 212 bp and a further domain that accounts for the difference in monomer size between species. Of this variable region, 35–50% consists of TTAGGG repeats (Adegoke *et al.* 1993). Cytologically these satellite sequences are found in terminal locations (Árnason *et al.* 1978; Árnason and Widegren 1989) and therefore much of the strong $(TTAGGG)_n$ hybridization in whale subtelomeric heterochromatin probably reflects these satellite sequences rather than large terminal arrays of uninterrupted $(TTAGGG)_n$.

A similar example of a terminal satellite containing TTAGGG repeats is found in the cat. A major satellite of this species consists of tandem repeats of a 483 bp underlying monomer unit, which in part consists of TTAGGG repeats. It is found at the terminal heterochromatin of some but not all telomeres, and as it is 1–2% of the total genome undoubtedly contributes to the $(TTAGGG)_n$ *in situ* signal at cat telomeres (Fanning 1987).

The interpretation of cross-hybridization approaches is further complicated by the unusual annealing properties of telomeric oligonucleotides *in vitro*, making it dangerous to predict how they will behave. For example, poly(CA) hybridizes to yeast TG_{1-3} repeats even at high stringency (e.g. Walmsley *et al.* 1984). In another example, Meyne *et al.* (1989) have documented the *in vitro* annealing properties of a variety of telomeric oligonucleotides, and some heterologous sequences anneal very well. For example, a *Tetrahymena* telomeric oligonucleotide, $(GGGGTT)_7$, annealed to a human $(TAACCC)_7$ oligonucleotide has a melting temperature of 65°C in 50 mM NaCl. In comparison the melting temperature of human-human $(GGGTTA)_7$-$(TAACCC)_7$ oligonucleotides is only slightly greater (70°C). Therefore it is possible that telomeric oligonucleotides will cross-hybridize to non-identical sequences even under normal high-stringency conditions. This behaviour may have led to the situation where $(TTGGGG)_n$ (Allshire *et al.* 1988), $(TTAGGG)_n$ (Moyzis *et al.* 1988), and $(TTTAGGG)_n$ (Richards and Ausubel 1988) were all reported to hybridize to human telomeres (either *in situ* or by detecting BAL31-sensitive signals on Southern blots). However, when human telomeres were cloned in yeast (Brown 1989; Cross *et al.* 1989) their sequence revealed the terminal repeat to be TTAGGG. Thus some care must be taken when using hybridization results alone to predict telomeric sequences.

Cloning telomeric DNA

Maxam–Gilbert chemical sequencing of genomic DNA has provided the terminal repeat sequence for a small number of species (e.g. *Euplotes*). Although dideoxy sequencing of genomic DNA has been successfully applied to the terminal repeats of tomato (Ganal *et al.* 1991) this requires a prior knowledge of the likely terminal repeat sequence so that a suitable sequencing primer can be used. In most species direct sequencing of genomic DNA to determine the terminal repeat sequence is impossible, and it is therefore necessary to clone telomeric sequences. A popular method is to use a modified yeast artificial chromosome (YAC) vector and to select for telomere function in the yeast *S. cerevisiae*. This is sometimes termed '1/2-YAC' telomere cloning. This was first done by Szostak and Blackburn (1982) who demonstrated that cloned *Tetrahymena* telomeres would allow a plasmid to replicate as a linear molecule when introduced into yeast. They then removed one *Tetrahymena* telomere and ligated random fragments of yeast genomic DNA. Following transformation into yeast and selection for transformants containing linear extrachromosomal plasmids they obtained a stable linear molecule with a *Tetrahymena* telomere on one end and an authentic yeast telomere on the other.

Why do heterologous sequences appear at first sight to function as telomeres in yeast? *Tetrahymena* terminal repeat sequences are not the same as those of yeast yet appear able to stabilize the ends of linear molecules in *S. cerevisiae*. The sequence of such a linear molecule being stably maintained in yeast, initially having terminated in heterologous repeats, can be determined by subcloning the terminal restriction fragments into a bacterial vector. This reveals that the heterologous (e.g. *Tetrahymena*) repeats have had yeast TG_{1-3} repeats added onto them. Therefore the heterologous repeats are not causing the addition of further copies of their own repeat sequence; rather they are seeding the addition of yeast repeats. It is important to stress that these linear molecules are stabilized by the function of the yeast telomeres, with the heterologous sequences acting to stabilize these linear molecules indirectly by seeding the addition of yeast TG_{1-3} repeats. One explanation for why these heterologous repeats should be able to seed the addition of yeast TG$_{1-3}$ repeats may be that the putative yeast telomerase, while adding yeast repeats, does not have a highly stringent substrate requirement. This is certainly the case for ciliate and human telomerase *in vitro*, which will add terminal repeats onto a range of TG-rich oligonucleotides (see Chapter 4). We should, however, note that telomerase has yet to be identified in yeast.

The terminal repeats from a wide range of species will stabilize linear molecules in yeast by seeding TG$_{1-3}$ addition, and this has enabled the successful cloning of telomeres from a range of organisms. Subcloning of the terminal restriction fragments from the resulting YACs into plasmid vectors has allowed analysis of the DNA sequence (e.g. Richards *et al.* 1992). In addition, telomeric sequences can be cloned directly into plasmid or phage vectors. The chromosome may not terminate in a blunt end, and to make it capable of being cloned into a vector it may be necessary to remove any single-stranded overhangs by, for example, pre-treating the genomic DNA with a nuclease such as BAL31 (e.g. de Lange *et al.* 1990). To enrich for terminal fragments a so-called 'end-clone' library can be made; this is described in detail in the section on *Arabidopsis* telomeres (Richards and Ausubel 1988; see below).

Criteria for a sequence being telomeric

A number of criteria can be used to help determine if a candidate sequence is indeed telomeric. Fulfilling just one of these criteria does not necessarily prove that the sequence is telomeric, which usually requires more than one line of evidence. Nevertheless, a feature such as a smeared band on a Southern blot, while not proof, has often been the first clue that a sequence is telomeric.

1. The signal detected on a Southern blot when the candidate sequence is used as a hybridization probe should be sensitive to pre-treatment of the intact genomic DNA with exonuclease BAL31. In contrast, a chromosome-internal sequence will be resistant to such digestion. A negative result does not prove the sequence is non-telomeric, as a number of species have large blocks of internal telomere-like repeats which mask the terminal signal.

2. The terminal restriction fragments of a chromosome are often heterogeneous in length. This reflects the dynamic turnover of terminal repeat sequences resulting from the processes of loss by incomplete replication and addition by, for example, telomerase. For example, as a result of this treadmilling the (TTAGGG)$_n$ signal detected on Southern blots of human DNA digested with an enzyme such as *Sau*3A is a smear. This

can be a useful first indication that a genomic fragment is terminal but it does not apply to all species; the macronuclear telomeres of some ciliates are extremely homogeneous in length (Oka *et al.* 1980; Klobutcher *et al.* 1981; Pluta *et al.* 1982).

3. The telomere is the end of the chromosome and as such can be detected by restriction mapping around the genomic location of the sequence. The telomere provides a pre-existing break and thus appears to be a site cleaved by all restriction enzymes, producing a characteristic cluster of restriction sites on a genomic map.

4. There should be no genetic markers on the linkage map of that chromosome which are more telomeric than a marker based on the candidate sequence. How stringent proof this is depends on the resolution of the genetic map.

5. The sequence should detect a terminal signal when used as a probe for *in situ* hybridization on metaphase chromosomes, although there may also be interstitial sites. A sequence required for essential telomere function would be expected to be present at all telomeres and also at any newly formed ones, such as those formed by the healing of breaks which can occur spontaneously or as the result of programmed genome rearrangements (see Chapter 6). The addition of terminal repeat sequences and not a subtelomeric sequence when healing a chromosomal break or capping an introduced linear plasmid, has often (e.g. *S. cerevisiae, Schizosaccharomyces pombe*, humans, and ciliates) provided good evidence that the terminal repeat sequence alone is sufficient for essential telomere function. In contrast there are many examples of subtelomeric sequences such as the yeast X and Y' elements that are found at only a subset of telomeres. For such sequences it can be difficult to determine if they do play a role in some aspect of telomere biology or whether they accumulate in telomeric regions for some unconnected reason.

6. Telomerase activity has been identified biochemically and shown to add the appropriate terminal repeat sequence in humans, mice, *Euplotes, Tetrahymena*, and *Oxytricha* (see Chapter 4). This provides strong evidence that a particular sequence is a terminal repeat. For example, (TTAGGG)$_n$ cross-hybridizes to sequences at mouse telomeres (Chapter 10) but the identification of a mouse telomerase which adds TTAGGG repeats (Chapter 4) strongly suggests that TTAGGG is indeed the mouse terminal repeat sequence.

General features of telomeric sequences

Terminal repeat arrays

Before detailing telomere structure by species, it is useful to consider some
of the general points that have emerged from their analysis. This section
therefore summarizes some features of the terminal repeat arrays, with
further details listed under the appropriate species. Subsequently general
features of subterminal sequences are reviewed.

   In most but not all cases the very end of a eukaryotic telomere is composed of a tandem repeat array of a short sequence. Most repeats conform
to a rather loose 'consensus' of $(A/T)_m(G/C)_n$. Some are $G+C$-rich but
this is by no means universal; the *Chlamydomonas* repeat, TTTTAGGG,
is $A+T$-rich. For many there is a strand bias in the distribution of G
residues, with the G-rich strand oriented 5' to 3' towards the end of the
chromosome; this strand is the presumed substrate for telomerase. There
are many examples cited below and in Chapter 6 of *de novo* telomere
formation. This may occur during programmed genome rearrangements,
the healing of a chromosome break, or telomere addition onto an introduced linear molecule. In almost all cases examples can be found where
only terminal repeat sequences are added, arguing strongly that they
are sufficient for essential telomere function. In some cases subterminal
sequences may be added as well, as seen in *S. cerevisiae* where healing can
be accompanied by the addition of both Y' and TG_{1-3} sequences. An
exception to the presence of short terminal repeats at chromosome ends is
found in *Drosophila* (Chapter 9). Table 3.1 details the sequences of terminal
repeats analysed to date. Only those that have been sequenced or have been
shown to be synthesized by telomerase from that particular species are
included.

Single-stranded nicks

There is some evidence to suggest that there are nicks in the terminal repeat
arrays of some species. This has usually been inferred from the ability of
DNA polymerase I to nick translate and label terminal restriction fragments of genomic DNA in the absence of DNase I. This is the case for
Tetrahymena (Blackburn and Gall 1978), *Physarum* (Johnson 1980),
Trypanosoma (Blackburn and Challoner 1984) and *S. cerevisiae* (Szostak
and Blackburn 1982). The pattern of nucleotide incorporation often indicates that there are nicks in both the C-and G-rich strands. The ability to
nick translate is frequently not inhibited by prior treatment with DNA
ligase. This could indicate a gap of one or more base pairs in the sequence
or nicks of a form that cannot be ligated by DNA ligase, perhaps for
example caused by a chemical reaction during the preparation of the DNA.
Although their exact nature remains speculative, one possibility is that

Table 3.1 Terminal repeat sequences

Sequence	Species
TTTTGGGG	*Euplotes, Oxytricha, Stylonychia*
TTGGGG	*Glaucoma, Tetrahymena*
TTGGGG and TTTGGG	*Paramecium*
TTTAGGG and TTCAGGG	*Plasmodium*
TTAGGG	*Trypanosoma brucei*
TAGGG	*Giardia lamblia*
$(TG)_{1-6}TG_{2-3}$	*Saccharomyces cerevisiae*
$TTACAG_{1-8}$	*Schizosaccharomyces pombe*
TTAGGGGG	*Cryptococcus neoformans*
TTAGGG	*Podospora anserina*
TTAGGG	*Fusarium oxysporum*
TTAGGG	*Neurospora crassa*
ACGGATGTCTAACTTCTTGGTGT	*Candida albicans*
TTAGGG	*Didymium iridis*
TTAGGG	*Physarum polycephalum*
AG_{1-8}	*Dictyostelium discoideum*
TTAGGC	*Ascaris lumbricoides*
TTGCA	*Parascaris univalens*
TTAGG	*Bombyx mori*
HeT-A, TART (retroposons)	*Drosophila melanogaster*
TTAGGG	Humans, mice
TTTAGGG	*Arabidopsis thaliana*
TT(T/A)AGGG	Tomato
TTTTAGGG	*Chlamydomonas reinhardtii*

Listed are those repeats that have been sequenced, either directly from genomic DNA or following cloning, or that are synthesized by telomerase from that particular species.

such nicks reflect incompletely synthesized Okazaki fragments, perhaps because DNA polymerase has some difficulty in replicating telomeric sequences.

Interpretation of such studies is complicated by the use of only a limited subset of the dNTPs in the nick translation reaction. For example, in *Trypanosoma brucei* DNA polymerase I gives significant incorporation into DNA by nick translation if radiolabelled dATP+dCTP+dTTP or dATP+dGTP+dTTP are included in the reaction (Blackburn and Challoner 1984). The other two combinations of three dNTPs do not give significant incorporation into total *T. brucei* genomic DNA. When analysed, the label has been incorporated into (TTAGGG)$_n$ tracts. In a similar fashion yeast telomeres preferentially label if only dGTP and dTTP are included in the reaction. However, in a relatively simple genome like *T. brucei* or yeast the terminal repeat arrays may be the only region of any

size with a strong base asymmetry between strands. Thus it is possible
to speculate that if there are nicks distributed at random throughout the
genome, if only three nucleotides are included in the reaction a nick will
only be extended for any distance if it occurs in a simple sequence arrays
with one strand containing only those three bases. Thus in *T. brucei* the
majority of the incorporation might occur in the (TTAGGG)$_n$ array
because most 'normal' genomic regions contain all four bases. Nicks are
inevitably introduced experimentally during DNA preparation, and indeed
if all four dNTPs are included in the *T. brucei* reaction there is enormous
incorporation of signal. Thus although there appear to be nicks in some
terminal repeat arrays, it remains unclear what fraction have been produced
during the preparation of the genomic DNA and indeed if nicks in the
terminal repeat arrays are any more frequent than nicks elsewhere in the
genome. One argument against a significant degree of nicking in either
strand of the terminal repeat arrays is the observation that the macronuclear
DNA of ciliates such as *Oxytricha*, *Glaucoma*, and *Stylonychia* can be end-
labelled and Maxam–Gilbert sequenced directly, irrespective of whether
the DNA is 3′ or 5′ labelled. These reactions produce fragments that can
extend past the terminal repeat arrays (Oka *et al*. 1980; Katzen *et al*. 1981;
Klobutcher *et al*. 1981; Pluta *et al*. 1982). Although the terminal repeat
arrays in these three ciliates are very short and it may not be relevant to
compare them with the much larger terminal repeat arrays in *S. cerevisiae*
or *T. brucei*, it remains unclear if specific nicks in terminal repeat arrays
do exist *in vivo* and if so what their biological role might be.

G-strand extension

Direct sequencing has shown that the G-rich strand is longer than its com-
plement in *Oxytricha*, *Stylonychia*, and *Euplotes*, thus producing a single-
stranded overhang. The reactivity of single-stranded thymidines to osmium
tetroxide indicates that there is also a G-strand overhang in *Tetrahymena*
and *Didymium* (Henderson and Blackburn 1989). However, it is not possible to determine if such an overhang exists for most species as this approach
requires the ability to sequence telomere sequences chemically using end-
labelled genomic DNA. If a single-strand overhang exists in *S. cerevisiae*
then for most of the cell cycle it is shorter than 30 nucleotides (Wellinger
et al. 1992, 1993*a,b*). An overhang corresponding to the 'primer gap' is
a predicted consequence of the end-replication problem and telomerase
action (see Fig. 1.5).

Irregularity in repeat arrays

The terminal repeats of some species, such as ciliates, are very homogeneous. However, there are other organisms that show a certain amount
of variation from repeat to repeat. For example, in *S. cerevisiae* the
repeats consist in general of single Ts followed by one to three Gs, and in

Dictyostelium they are composed of single As followed by one to eight Gs. *Schizosaccharomyces pombe* also has rather variable repeats. In theory some variability could reflect the accumulation of mutations. Incomplete replication and telomerase-mediated sequence addition are processes confined to the tip of the chromosome. Thus there is predicted to be a rapid turnover of sequences at the very end of the chromosome. In contrast, more internal repeats may be turned over only infrequently. Examples of greater sequence turnover in the more terminal regions of a repeat array have been shown for *Plasmodium berghei* and *S. cerevisiae* (see below). If during the evolution of a species such a sequence is rarely if ever turned over it could accumulate mutations, and this in turn could lead to apparent sequence heterogeneity in a terminal repeat sequence. For example, the internal $(TTAGGG)_n$ array found on human chromosome 2 is very degenerate in sequence, with only about half of the repeats being exactly TTAGGG (Wells *et al.* 1990). The array is believed to result from the fusion of two great ape chromosomes to form the human chromosome 2 a few million years ago and, as it is no longer subject to turnover by telomerase and incomplete replication, the mutations that have accumulated have not been removed. In contrast, the terminal regions of telomeric human $(TTAGGG)_n$ arrays are very homogeneous in sequence, although they do become rather degenerate in the region immediately adjacent to the subtelomeric sequences (Allshire *et al.* 1989; Brown *et al.* 1990; de Lange *et al.* 1990). It is therefore possible to speculate that $(TTAGGG)_n$ arrays may become variable if not subject to constant loss by incomplete replication and then *de novo* synthesis by telomerase.

One might argue that accumulation of random mutations is unlikely to account for the variation seen in *S. cerevisiae* for a number of reasons. Firstly, random point mutations would be expected to introduce a range of different base-pair changes, whereas the yeast arrays are composed solely of Ts and Gs; a similar argument can be applied to *Dictyostelium*. In both species almost all the sequence variation is accounted for by variation in the number of guanine residues. Secondly, sequence extending almost to the very last base of a number of terminal repeat arrays in yeast has been obtained and the variation continues right up to the end. Thirdly, newly formed telomere arrays produced by healing of heterologous telomeres in yeast are variable throughout the length of their TG_{1-3} arrays. Both the pattern of variation within yeast and *Dictyostelium* terminal repeat arrays and the rapidity with which variation occurs are inconsistent with the observed variation being the result of simple point mutational changes. We must therefore seek other explanations for the generation of sequence variation at yeast telomeres. One possibility comes from the observation that certain mutant telomerase RNAs promote slippage on the template during synthesis, leading to variation in the number of Gs in each repeat (Chapter 4). Although little more than speculation at this stage, it is possible that the variation in the number of Gs in *S. cerevisiae* and *Dictyostelium* might reflect slippage synthesis by telomerase, a hypothesis which could be tested if telomerase were to be isolated from either species.

Interspersion

Some species appear to have arrays that are composed of two similar repeats. In *Paramecium* these are TTGGGG and TTTGGG, and in *Plasmodium* TTTAGGG and TTCAGGG. The two terminal repeats appear to be randomly interspersed with each other. Tomato telomeres also consist of a mixture of terminal repeats, in this case of TTTAGGG and TTAAGGG, but it is not known if the interspersion is random. One speculation is that the presence of two variant repeats reflects the presence of two types of telomerase activity in the cell, one for each repeat. For example, there could be two genes for the telomerase RNA component, differing by a single base in the template region. Because telomerase activity has not been identified in any of these species it is not possible to test this hypothesis as yet.

Terminal repeat array lengths vary from species to species

The length of a repeat array varies in two ways. Firstly, it can vary between species, from 20 bp in *Oxytricha* macronuclei to over 100 kb in mice. Secondly, the length of a specific terminal repeat array is not always absolutely fixed as it is in *Oxytricha*. In many other species the amount of terminal repeat sequence at any particular chromosome end varies between cells in a population, producing a characteristic smeared signal for terminal restriction fragments as detected by Southern analysis. This reflects the dynamic nature of telomeric sequences, subject to continual loss and addition processes in dividing cells. A detailed examination of how a single yeast cell gives rise by division to a population of cells with heterogeneity in terminal repeat array sizes can be found in Shampay and Blackburn (1988). The telomeres of replicative descendants of each individual become progressively more heterogeneous in length with each round of cell division, and a model where telomeres are continually being lengthened and shortened by small increments fits these *in vivo* data well.

Internal tracts

Telomere repeats are not always found only at chromosome ends. They can be found at internal chromosome locations in a range of species including *S. cerevisiae*, *Tetrahymena*, *Oxytricha*, *Stylonychia*, and *Paramecium*. Indeed, in *Physarum* most of the terminal repeat-like sequence is chromosome-internal and therefore BAL31-insensitive. A number of vertebrates show strong $(TTAGGG)_n$ hybridization in regions of constitutive heterochromatin, such as the dramatic pericentric blocks of degenerate $(TTAGGG)_n$ repeats in the guinea pig, or the whalebone whale satellite that has within its repeat unit a short (TTAGGG)$_n$ tract. Another example is *Arabidopsis thaliana*, which has degenerate telomere-like repeats present within a repeated sequence found at its centromeres (Simoens *et al*. 1988; Richards *et al*. 1991). It is not clear in most species if these internal repeats have any role in telomere function. They may reflect some aspects of karyotype evolution, such as telomere–telomere fusions. Examples of the relationship of internal telomere tracts to ancestral telomere fusions are discussed in Chapter 10.

G + C content

It is often stated that terminal repeats are G + C-rich, whereas in reality many are less than 50% G + C. For example, the *Chlamydomonas* repeat is (TTTTAGGG)$_n$ and is less than 40% G + C. A more accurate statement is that there is often a bias in base composition so that one strand is G-rich. Of those species with a short repeat that have been analysed to date a strand bias has been found for all except *Parascaris*, and for those species with a strand bias the G-strand runs 5′ to 3′ towards the terminus of the chromosome. There are a number of possible explanations for this. Firstly, only one strand is synthesized by telomerase and there may be an as yet unknown biochemical feature of the reaction that prevents telomerase from efficiently synthesizing sequences which are not G-rich (Blackburn 1992). Secondly, the G-strand is the one that will form the single-stranded terminal overhang. The combination of being single-stranded and G-rich leads to the potential to form novel structures stabilized by G·G bonds (see below). If such structures have a role in telomere function this could provide an explanation for the evolutionary conservation of this G-strand bias.

A feature that has been noted by Petracek *et al*. (1990) is that the G + C content of terminal repeat sequences is often inversely related to the overall G + C content of the genome (Table 3.2). There is little correlation between the two figures for genomes with G + C content in the 35–55% range but there is a marked effect for species with either a very A + T- or G + C-rich genome. As illustrated in Table 3.2, *Chlamydomonas* has an extremely G + C-rich genome and one of the most A + T-rich telomere sequences, whereas *Tetrahymena* and *Dictyostelium* have very A + T-rich genomes and have the most G + C-rich of all the terminal repeat sequences.

The reason for this bias in terminal repeat base composition is not known. One could speculate that it would benefit a cell to minimize the number of sequences similar to the terminal repeat in its genome because this would reduce the chance of the sequence at a random break being similar enough to a terminal repeat to be recognized by telomerase and healed, causing a karyotypic change. As illustrated in Fig. 3.1, for species with a genomic G + C content of around 50%, the base composition of the terminal repeat sequence has little effect on the frequency of its occurrence at random in the genome. However, *Chlamydomonas, Dictyostelium,* and

Table 3.2 Frequency of occurrence of an 8 bp non-palindromic sequence in genomes of different G+C content

% G+C		Frequency of random matches[a] (kb)
Genome	Terminal repeat	
25	25	11
	50	104
	75	932
50	25	33
	50	33
	75	33
75	25	932
	50	104
	75	11

[a] Average spacing between random occurrences of an 8 bp non-palindromic sequence.

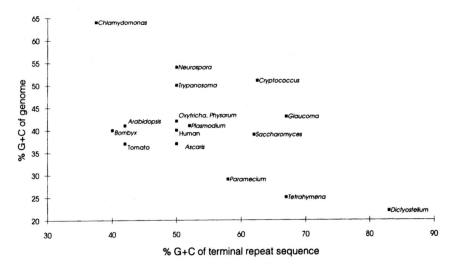

Fig. 3.1 Relative % G+C of telomere repeats and genomic DNA for various species. The % G+C of the genome of each species is plotted against the % G+C of its terminal repeat sequence. Values are taken from Petracek *et al.* (1990) and references therein and Fasman (1976).

Tetrahymena may have taken advantage of their genome composition by having terminal repeat sequences with a reverse bias in $G + C$ content. The result is that they have reduced the number of random telomere-like sequences in their genomes. However, whether the advantages of such a situation indeed explains the inverse correlation between genome and terminal repeat sequence base composition in these three species is not clear.

Subterminal sequences

In some cases there is a direct transition between the terminal repeat arrays and the unique internal sequences. However, in many species the two are separated by repetitive sequences which are often species-specific, with a variety of lengths and degrees of repetitiveness. Interest in them stems from the fact that many are found only at telomeres, and could in principle have a role in telomere function. In most cases they may not have an essential role, as often they are present only at a subset of telomeres, but this does not preclude some other role not required at every telomere.

Even if most subtelomeric repeats do not have an essential role in telomere biology, they could still be implicated in a number of phenomena. For example, they could facilitate chromosome pairing in meiosis (Chapter 2). The presence of homologous sequences at the telomeres of non-homologous chromosomes could lead to telomere exchange by recombination between the chromosomes, as is suggested to occur in humans (Chapter 10) and produces variation in the size of *Plasmodium* chromosomes (see below). Telomerically located genes, such as the *MAL, PHO*, and *SUC* genes in *S. cerevisiae* and the VSG genes in trypanosomes, may evolve and move around the genome by a similar mechanism. Subterminal repeats could also provide a buffer zone between essential genes and the terminus to protect them from the dynamic processes of loss and addition occurring at the terminal repeat arrays. They could also buffer genes against telomere position effects on gene expression (Chapter 8) although the Y' element in fact seems to increase the extent of spreading of telomeric position effect in yeast (Renauld *et al.* 1993). Another possibility is that they could facilitate the healing of a chromosome that has lost its terminal repeat arrays by telomere conversion between a subterminal repeat and one on another chromosome, as can occur in *S. cerevisiae* (Dunn *et al.* 1984) and trypanosomes (de Lange *et al.* 1983). Finally, a number of the subterminal repeats form large blocks of constitutive heterochromatin with the repeats forming a significant fraction of the total genome. The role of these heterochromatic blocks, which are not found at all higher eukaryotic telomeres, is still very unclear and may not be related to telomere function.

A converse speculation is that subtelomeric sequences do not confer any benefit on telomere function and have not been selected for by a role at telomeres. Rather they might be present at telomeres for a mechanistic reason reflecting a different behaviour of these regions to elsewhere in the genome. For example, a reciprocal crossover between two non-homologous chromosomes in meiosis will lead to deletion or duplication of chromosomal regions in offspring. The further internal the translocation sites are the greater the extent of the deleted or duplicated region, thus increasing the likelihood that such a rearrangement will be lethal to the offspring. In contrast, recombination between two subtelomeric regions will not alter the location of any essential genes as these are likely to be internal to the breakpoint. As a consequence of this, recombination and exchange may appear to occur more often between telomeres than elsewhere in the genome because the translocation chromosomes can be segregated to yield viable offspring. If a repetitive sequence evolves as a side-effect of recombination between chromosomes it may therefore arise preferentially at a location which is permissive for such recombination events. It is possible to speculate that repetitive sequences may tend to arise in subtelomeric locations, where recombination between non-homologous chromosomes is permitted, rather than elsewhere in the genome. Once a repetitive sequence has arisen telomere exchange can also facilitate its dispersion to other telomeres. Such behaviour is one speculative explanation for the clustering of VNTR minisatellite loci in the terminal regions of the human genome (Chapter 10) and the evolution of terminal satellite blocks. An example of recombinational expansion of an array of subtelomeric Y' elements is detailed in Chapter 9.

In summary it is very unclear if subterminal sequences have any role in telomere function, or whether they have arisen as a by-product of an intrinsic feature of these regions of the genome. Most subterminal repeats are sequences in search of a function.

Telomeric sequences

The following section summarizes the main points of what is known of telomeric sequences, both terminal repeats and subtelomeric sequences, in a wide variety of species.

Lower eukaryotes

Oxytricha, Stylonychia, and Euplotes

The macronuclear telomeres (see Chapter 6) of these hypotrichous ciliates can be chemically sequenced following either 3' or 5' end-labelling (Oka *et al.* 1980; Klobutcher *et al.* 1981; Pluta *et al.* 1982) and are composed of the sequence $(TTTTGGGG)_n$. These species maintain a very precise terminal repeat structure, with all telomeres having the same length and sequence. For *Oxytricha* and *Stylonychia* there is 20 bp of duplex DNA with 16 bp of single-stranded overhang:

GGGGTTTTGGGGTTTTGGGGTTTTGGGGTTTTGGGG-3'
CCCCAAAACCCCAAAACCCC-5'

Euplotes has a longer duplex region and the overhang is two bases shorter, finishing with two rather than four Gs:

GGGGTTTTGGGGTTTTGGGGTTTTGGGGTTTTGGGGTTTTGG-3'
CCCCAAAACCCCAAAACCCCAAAACCCC-5'

Most of the (TTTTGGGG)$_n$ repeats in the *Oxytricha* micronucleus are BAL31-sensitive and thus terminal (Dawson and Herrick 1984). Interestingly, the blocks of (TTTTGGGG)$_n$ at the micronuclear telomeres are many kilobases in length, much longer than the arrays at macronuclear telomeres. This suggests that whatever process produces the precise, short macronuclear telomeres is operating differently in the micronucleus. A specific telomere processing reaction which produces these defined terminal structures in the macronucleus has been described in *Euplotes* (see Chapter 4).

Glaucoma

This holotrichous ciliate has (TTGGGG)$_n$ terminal repeats. This has been shown (Katzen *et al.* 1981) by end-labelling macronuclear DNA and chemical sequencing the C-strand, which again terminates at a constant position:

... AACCCCAACCCCAACCCCAAC-5'

The G-strand has not been sequenced directly and it is not known if there is a single-stranded overhang.

Tetrahymena

Biochemical analysis 25 years ago of the macronuclear genome of *Tetrahymena* revealed (TTGGGG)$_n$ repeats at the ends of the extrachromosomal rDNA molecules (Blackburn and Gall 1978) which were later shown to be BAL31-sensitive (Yao and Yao 1981). Since then telomeres from a number of *Tetrahymena* species have been cloned and sequenced; and all terminate in (TTGGGG)$_n$ (Challoner and Blackburn 1986).

There are also a large number of interstitial (TTGGGG)$_n$ arrays in the *Tetrahymena* micronuclear genome. These are short and all appear to flank a dispersed repetitive element with the structural hallmarks of a transposable element (Cherry and Blackburn 1985). A similar class of telomere-bearing transposons has also been identified in *Oxytricha fallax*, in this case flanked by (TTTTGGGG)$_n$ repeats (Herrick *et al.* 1985; Williams *et al.* 1993). These transposon-like elements are found only in the micronucleus and are precisely excised as circles during macronuclear development (Chapter 6). Indeed, it has been suggested that all internally eliminated sequences are related to transposons and are excised by some form of transposase-mediated mechanism (Hunter *et al*. 1989). *Stylonychia* also has many short internal telomeric repeats in its micronuclear genome, but these are found within a conserved element with few of the transposon-like features of the elements described in *Oxytricha* (Stoll *et al*. 1991, 1993).

Paramecium

Terminal repeat arrays have been cloned and sequenced and consist of a highly interspersed mixture of two main motifs, TTGGGG and TTTGGG. Many of these data come from studying the process of *de novo* telomere addition. As discussed in detail in Chapter 6, *Paramecium* appears to be able to add terminal repeat sequences onto almost any sequence *in vivo*, ranging from injected bacterial plasmid DNA to breaks that occur during genome rearrangement. One implication is that *Paramecium* telomerase may have a very lax substrate requirement *in vivo*.

Linear DNA from probably any source is replicated up to a very high copy number when injected into *Paramecium*. The presence of *Paramecium* telomeric sequences at the ends of the injected molecule inhibits the end-to-end fusions that occur with molecules lacking telomeric sequences (Bourgain and Katinka 1991).

There are interstitial $(TT[T/G]GGG)_n$ repeats in the *Paramecium* genome. Curiously these appear to be hotspots for illegitimate recombination. Katinka and Bourgain (1992) have analysed the sites of integration of the yeast *HIS3* gene into the *P. primaurelia* genome. From the frequency at which they could obtain genomic clones containing both the integrated *HIS3* gene and a repeat sequence array they estimate that 30% of integrants are within or close to an interstitial $(TT[T/G]GGG)_n$ array. This is supported by sequence analysis of a small number of integration sites, some of which contain the *HIS3* gene within or very close to an interstitial telomere repeat array. However, if integration occurred at random in the genome this would be expected to occur at a frequency of less than 5×10^{-4} because there are only some 25–60 such arrays in the *Paramecium* genome. It therefore appears that these arrays are hotspots for integration, but the underlying reason is a mystery.

Plasmodium berghei

The terminal repeat arrays of this species of protozoan parasite have been cloned and sequenced and consist of a random mix of TTTAGGG and TTCAGGG repeats (Ponzi *et al*. 1985). Taking advantage of the pattern of interspersion the rate of turnover of the terminal repeat arrays has been estimated by cloning and sequencing the same telomere repeatedly from a *Plasmodium* population. All had an identical interspersion pattern in the more internal region, but the clones became more varied as the end of the telomere was approached. This would be consistent with a model where there is a high rate of turnover of repeat sequences at the very end of the molecule and progressively less turnover further in. Such a dynamic flux is clearly consistent with a telomerase model of telomere replication and is evidence that new repeats are added to the ends of telomeres *in vivo* (Ponzi *et al.* 1992).

Plasmodium chromosomes are highly polymorphic in size and often vary from strain to strain. Many chromosomes have large arrays of a 2.3 kb sequence subtelomerically. Each 2.3 kb monomer contains a 160 bp stretch of terminal repeat sequence (Dore *et al.* 1990). In one example a 400 kb difference in the size of chromosome 7 between two strains can be accounted for by differences in the number of copies of this 2.3 kb sequence, which is probably due to either recombination between non-homologous chromosomes or unequal crossover between homologues (Ponzi *et al.* 1990).

Another example of chromosome size polymorphism again results from differences in the copy number of this 2.3 kb repeat, although in this example the shorter telomere completely lacks the 2.3 kb sequence (Pace *et al.* 1990). The 50 kb increase in size of the shorter telomere reflects the acquisition of multiple copies of the 2.3 kb sequence. It has been suggested that this was produced by recombination of the terminal repeat array on this chromosome with a similar tract present within one of the 2.3 kb elements in a subtelomeric array of a non-homologous chromosome. In conclusion much of the chromosome size polymorphism that is characteristic of *Plasmodium* may be explained by recombinational exchanges involving telomeric and subtelomeric repetitive sequences.

Plasmodium falciparum

Telomeric sequences have been cloned both in plasmid vectors and using 1/2-YAC telomere cloning in yeast (Vernick and McCutchan 1988; de Bruin *et al.* 1992). The terminal repeats have the same consensus sequence, TT[T/C]AGGG, as *P. berghei* (Ponzi *et al.* 1985). The terminal repeat arrays exhibit genetic hypervariability. A probe for the terminal (TT[T/C] AGGG)$_n$ repeats detects many bands on blots of digested DNA run on conventional gels (Vernick *et al.* 1988). A large percentage of the bands seen in the progeny of a cross do not correspond to alleles present in either parent. Restriction fragments apparently at internal sites very close to the terminus can also vary. This argues that this variation is not because of turnover of terminal repeat sequences, as this would be unlikely to change the size of an internal fragment while still maintaining the restriction sites flanking both sides of it. One speculation is that this variation reflects similar processes to those that cause the formation of new VNTR alleles in humans (Chapter 10).

As with *P. berghei*, homologous chromosomes in independent isolates frequently vary in size in a dramatic fashion. The karyotype, visualized by pulsed field gel electrophoresis (PFGE), is largely stable through asexual propagation, although rare variants can arise (Corcoran *et al.* 1988). However, when two *P. falciparum* lines are crossed, progeny with chromosomes different in size from either parent readily occur (Walliker *et al.* 1987). The variation appears to arise mainly during meiosis. The restriction maps of chromosomes 1 and 2 in six cloned lines of *P. falciparum* indicate that the length variation is because of differences in the terminal regions. These contain a variety of repetitive subtelomeric sequences including a 21 bp element called rep20, and it is proposed that recombination between these sequences underlies the chromosome size variation in a similar fashion to what is proposed to occur in *P. berghei* (Corcoran *et al.* 1988; Vernick and McCutchan 1988). For a review of *Plasmodium* telomere structure and the recombination events underlying chromosome size polymorphisms see Foote and Kemp (1989).

Trypanosoma brucei

The terminal repeat sequence is $(TTAGGG)_n$ (Blackburn and Challoner 1984; Van der Ploeg *et al.* 1984). This was shown by sequencing clones from an end-library, and almost all the signal on blots is BAL31-sensitive.

The *T. brucei* genome consists of about 20 large (0.2–5.7 Mb) chromosomes plus around 100 minichromosomes (50–150 kb). It is similar to many other species in that moderately reiterated sequences are found subtelomerically at a number of telomeres. This is illustrated by the sequence of two minichromosome telomeres (Weiden *et al.* 1991). A $(TTAGGG)_n$ terminal repeat array is followed by repeats of a 29 bp TTAGGG-like sequence, a region containing 74 bp G + C-rich repeats separated by about 155 bp stretches of A + T-rich sequence, and then 50–150 kb of the 177 bp repeat which constitutes over 90% of the length of most, if not all, minichromosomes. This 177 bp repeat sequence is also present on some but not all of the larger chromosomes. A probe containing both the 74 bp G + C-rich repeats and the flanking A + T-rich sequence hybridizes to all chromosomes, separated by PFGE, and most of the signal detected by this probe is BAL31-sensitive. This is consistent with this G + C-rich sequence being present at all telomeres. It is unknown if it has any essential role in telomere biology in this species.

Giardia lamblia

Telomeric sequences cloned from this flagellated protozoan parasite terminate in $(TAGGG)_n$ repeat arrays (Adam *et al.* 1991; le Blancq *et al.* 1991). A $(TAGGG)_n$ probe detects an almost entirely BAL31-sensitive signal on Southern blots, indicating that this sequence is not found in substantial amounts at internal locations.

Saccharomyces cerevisiae

Yeast telomeres were first cloned by 1/2-YAC telomere cloning using a *Tetrahymena* telomere at one end of the molecule (Szostak and Blackburn 1982). When the *Tetrahymena* telomeres were subcloned into bacterial vectors and sequenced it was found that they had been capped by terminal repeat sequences which were subsequently found on authentic yeast chromosomes (Shampay *et al.* 1984; Walmsley *et al.* 1984). Cloning terminal repeats into plasmid vectors often utilizes BAL31 exonuclease to produce a blunt, ligatable end. However, such exonuclease digestion can lead to the loss of the most terminal few hundred base pairs. One way to overcome this problem is by the use of T4 DNA polymerase to produce a flush end. In *S. cerevisiae* this has enabled the cloning of sequences almost at the very end of the chromosome (Wang and Zakian 1990). The terminal repeats conform to a consensus $(TG)_{1-6}TG_{2-3}$. However, this is usually abbreviated TG_{1-3} or $C_{1-3}A$. The repeats are heterogeneous as illustrated by this example (Wang and Zakian 1990):

... TGTGTGGGTGTGGTGTGTGTGGGTGTGGGTGTGTGTGGGTGTGGGTGTGGTGTGTGG-3'

If the same telomere is cloned a number of times this reveals that the distal portion of the TG_{1-3} array varies from clone to clone whereas the more internal region is constant. This result provides evidence for a higher turnover rate of sequences at the end of the chromosome (Wang and Zakian 1990).

The ability to confer chromosome stability by the addition of TG_{1-3} repeats alone suggests that they are sufficient for essential telomere function. One example where a linear molecule is healed by the addition of yeast telomeric sequences is reported by Kämper *et al.* (1989). The yeast *Kluyveromyces lactis* contains what is termed a killer plasmid that can be transferred to and maintained in *S. cerevisiae*. It replicates as a linear plasmid in the *S. cerevisiae* cytoplasm and causes the secretion of a toxin; hence the 'killer' phenotype. The terminal sequences of the killer plasmid do not function as telomeres in the yeast nucleus. However, it is possible to force such a plasmid to replicate in the *S. cerevisiae* nucleus by integrating a copy of the *S. cerevisiae LEU2* gene into it *in vivo*; the *LEU2* gene can only be expressed when in the nucleus. Linear versions were obtained which had been healed by the acquisition of yeast telomeres. Yeast TG_{1-3} repeats were added directly onto the termini of the killer plasmid. No sequences other than TG_{1-3} are added and this example illlustrates healing without concomitant transfer of X or Y' elements (see below). In general, linear molecules introduced into the cell do not get healed by the addition of yeast telomeres. In this case, however, the healing may have been facilitated by the killer plasmid terminating in ... TGTGT-3', which is similar to the yeast terminal repeat sequence.

Saccharomyces cerevisiae has two main types of subtelomeric repeat, called X and Y'. Y' elements are found only at telomeres (Szostak and Blackburn 1982; Chan and Tye 1983*a*; Walmsley *et al.* 1984). They are highly conserved sequences, existing in long (6.7 kb) and short (5.2 kb) forms, the difference in length resulting from a deletion. Within the Y' sequence is a tandem array of 8–20 copies of a 36 bp repeat sequence (Horowitz and Haber 1984; Louis and Haber 1992). Y' elements are found immediately adjacent to the terminal TG_{1-3} repeats in 0–4 copies (Szostak and Blackburn 1982; Chan and Tye 1983*a*; Walmsley *et al.* 1984). There is a very striking variation in the copy number and location of Y' elements in *S. cerevisiae* strains (Chan and Tye 1983*a*; Horowitz and Haber 1984; Zakian and Blanton 1988; Louis and Haber 1990*a*). Y' elements are limited to *S. cerevisiae* and are not found in other *Saccharomyces* species, suggesting that they are evolving rapidly (Jäger and Philippsen 1989*a*). Further evidence that it is the TG_{1-3} repeats and not the Y' elements that are required for minimal telomere function is that Y' elements are not necessarily found on all chromosomes in a strain; some chromosomes lack them (Button and Astell 1986). This is readily visualized by probing a blot of intact yeast chromosomes separated by PFGE with a Y' probe; not all bands hybridize (for example see Jäger and Philippsen 1989*a*).

Between Y' and the rest of the chromosome arm is a less highly conserved repetitive sequence family called X (Chan and Tye 1983*a,b*). X and Y's are separated by internal arrays of TG_{1-3} repeats (Walmsley *et al.* 1984). Some chromosomes, as visualized by PFGE, do not carry X elements, and some lack both X and Y' (Zakian and Blanton 1988). However, a chromosome III lacking both X and Y' elements behaves normally both in mitosis and meiosis, suggesting neither element has an essential role in telomere function (Murray and Szostak 1986).

Y's are dynamic genomic elements. They can recombine with each other mitotically at a rate of around 10^{-6} per cell per generation, a rate similar to that seen for other repeated sequences in yeast (Louis and Haber 1990*b*). They can be transferred to a chromosome end that previously lacked a Y' element, and conversely recombinational events can leave a telomere Y'- free. Y' elements can move to a new location by both mitotic and meiotic recombination (Dunn *et al.* 1984; Horowitz *et al.* 1984; Louis and Haber 1990*b*). Y's can also exist as autonomously replicating circles (Horowitz and Haber 1985) perhaps as a result of intra-array recombination releasing a circular Y' element which can then replicate extrachromosomally; Y' elements contain an autonomously replicating sequence (ARS) which may provide a replication origin for the extrachromosomal circle (Chan and Tye 1983*b*). Such circles could then integrate into another Y' element at a different locus. These properties are likely to contribute to the generation of the strain differences that are seen with respect to copy number and location of Y' elements. The Y' sequence (Louis and Haber 1992) reveals two overlapping open reading frames, one of which encodes a polypeptide with homology to RNA helicases. The structure of Y' is consistent with it being of viral or transposable element origin. However, transposition cannot account for the majority of changes in Y' location which have been observed, although they are consistent with a recombinational mechanism. Y' elements have an unknown function but the expansion of single copies into tandem arrays has been observed, and thus they provide a model system for the evolution of the tandem repeat arrays which constitute the telomeric heterochromatin of many species, such as those of whalebone whales and various plant species described elsewhere in this chapter. The recombinational behaviour of Y' elements is discussed further in Chapter 9 in the context of a back-up pathway of telomere maintenance seen in *est1* survivors.

Y' elements can be associated with chromosome healing events. A linear plasmid terminating within a cloned Y' element can acquire a full-length Y' element and its TG_{1-3} cap by *RAD52*-dependent recombination with a Y' element at another telomere (Dunn *et al*. 1984). *RAD52* is required for most types of mitotic recombination in yeast, and in this case the cell may be taking advantage of a fairly conventional gene conversion event to heal a break. This contrasts with the *RAD52*-independent transfer of TG_{1-3} repeats onto heterologous terminal repeat sequences (Chapter 6). Another example of the potential for Y' elements to heal breaks comes from the study of a dicentric chromosome II made by integrating a second *CEN* element into one of the chromosome arms (Jäger and Philippsen 1989*b*). Only rarely are viable transformants obtained when attempting to perform this integration event. These turn out either to have one of the two *CEN* elements deleted or now to possess a chromosome II that is broken between the *CEN* elements to yield two monocentric chromosomes. As each of the breakage products is stable, telomeres must have been acquired at the break point. The original chromosome II did not carry Y' elements, but some of the broken chromosomes do. Although the exact molecular nature of the healing events is not known they are consistent with the broken end acquiring TG_{1-3} repeats by a mechanism involving Y' movement.

Schizosaccharomyces pombe

The terminal repeats have been cloned and sequenced from fission yeast (Sugawara and Szostak 1986; Sugawara 1989). Although overall they conform to a rather degenerate consensus ($TTACA_{0-1}C_{0-1}G_{1-8}$) over two-thirds of the repeats are of the form $TTACAG_{1-8}$. The majority of the remaining repeats are TTACACG, TTACACGG, and TTACGG. In this respect the repeat motif is somewhat similar in style to those in *Dictyostelium* and *Saccharomyces*, where the repeat variation is caused by differences in the number of guanines.

Evidence that these repeats are sufficient for essential telomere function

comes from studying the healing of chromosome breaks. Matsumoto *et al.*
(1987) have analysed the healing of chromosome breaks in *S. pombe* caused
by γ-irradiation. The data are consistent with only *S. pombe* terminal
repeat sequences having been added directly to the breakpoint; none of the
S. pombe subtelomeric sequences that have been identified (Sugawara
1989) are found at these new telomeres, suggesting they are not required
for essential telomere function. The healing event may be analogous to the
human α-thalassaemia example described in Chapter 6 where $(TTAGGG)_n$
repeats have been added directly onto a breakpoint.

Cryptococcus neoformans

This is a fungal parasite of humans. If a *C. neoformans* strain carrying a
ura5 mutation is transformed with the cloned *C. neoformans URA5* gene
as a linear molecule, some of the resulting Ura$^+$ transformants are stable
and have the gene integrated, whereas others are unstable and maintain
the introduced DNA as an extrachromosomal molecule. All the extrachromosomal molecules are linear, as shown by exonuclease sensitivity.
When the ends were cloned and sequenced they were found to have acquired
$(TTAGGGGG)_n$ repeats (Edman 1992).

A curious feature of this species, along with *Podospora anserina* and
Fusarium oxysporum (see below), is the different behaviour of linear and
circular molecules. If a circular plasmid carrying cloned telomeric repeats
is used the transformation frequency is quite low (*c.* 200 per μg). In contrast, if the same molecule is linearized by a restriction enzyme to free the
terminal repeats it now transforms with high (*c.* 10^5 per μg) efficiency.
Furthermore, the circular molecule is maintained only as an extrachromosomal element if it is rearranged to give a linear form. A circular molecule,
despite carrying all the same DNA sequences as the linear, cannot be maintained in this form extrachromosomally and must either integrate or
rearrange. This is in marked contrast to the situation in *S. cerevisiae* where
there is no difficulty in maintaining extrachromosomal circles. Two
possibilities immediately suggest themselves to explain the situation in *C.
neoformans*. Firstly the circular molecule may be incapable of replication
because $(TTAGGGGG)_n$ repeats act as entry points for replication complexes but only when terminating a linear molecule. This could occur if the
terminal repeats were recognized by a protein that could not bind them
when they were internally located. Another possibility is that circular
molecules are replicated but cannot be segregated properly. Replication of
a circle will lead to the two daughter molecules being catenated with each
other, and unless efficiently resolved (e.g. by topoisomerase II) they would
be unable to segregate from each other. Despite these speculations, the
reason for the inability to maintain circular molecules remains unknown.

Podospora anserina and Fusarium oxysporum

Tetrahymema telomeres will function in this filamentous fungus to permit the extrachromosomal maintenance of linear molecules by being capped by *Podospora* repeats (Perrot *et al.* 1987). Such healed ends have been cloned and sequenced and reveal the *Podospora* telomere repeat to be $(TTAGGG)_n$, with all the signal being BAL31-sensitive (Javerzat *et al.* 1993). The ends of an introduced molecule that has been rearranged to form an extrachromosomal linear molecule capped by *Fusarium* telomere sequences have been cloned and sequenced and are composed of $(TTAGGG)_n$ repeats (Powell and Kistler 1990). These sequences are also BAL31-sensitive in the genome. As with *C. neoformans*, circular molecules cannot be maintained extrachromosomally in either species; they either integrate or are rearranged to form linear molecules capped by telomeric sequences.

Neurospora crassa

One particular telomere (linkage group V right) has been cloned by walking from a locus that was known to be telomere-adjacent (Schechtman 1990). It terminates in $(TTAGGG)_n$ and most of the $(TTAGGG)_n$ in the genome is BAL31-sensitive. Additional telomeres have been isolated by screening an *N. crassa* library with a terminal repeat probe.

Candida albicans

This yeast has very unusual terminal repeats. Native telomeres and one example of a healed chromosome break have been cloned and sequenced. They terminate in 15–30 copies of a 23 bp sequence, 5'-ACGGATGTC-TAACTTCTTGGTGT-3', with the 3' end nearer the terminus of the chromosome (Sadhu *et al.* 1991; McEachern and Hicks 1993). This 23 bp sequence appears to terminate the chromosomes as judged by all molecular data available. The repeats are BAL31-sensitive and appear as diffuse bands on blots, suggesting length heterogeneity. The repeats are present on every chromosome, and compose the most distal nucleotides of every genomic clone containing the repeats, including a newly formed healed end. Furthermore, telomere length increases when cultures are grown at high temperatures and this growth involves increases in the number of these repeats. It cannot be ruled out that there are very short stretches of a more canonical repeat at the very end of the chromosomes, but one would have to say they are either extremely short and not readily cloned or are deleted from the clones at high frequency. In addition, *S. cerevisiae* and *Tetrahymena* repeats do not hybridize to *Candida* telomeric fragments even under low stringency conditions. This all points to this unusual sequence being the real terminal repeat sequence.

This repeat clearly does not conform to the simple telomere consensus, although the final 7 bp is like a *S. cerevisiae* TG_{1-3} repeat. It will be of great interest to see how widespread such an unusual terminal structure is in the yeasts and to discover if these larger repeats are synthesised by a telomerase enzyme or whether they are maintained by a telomerase-independent pathway.

Didymium iridis and Physarum polycephalum

These are both acellular slime moulds. Extrachromosomal rDNA telomeres from *Didymium* have been cloned and terminate in $(TTAGGG)_n$ arrays, and direct Maxam–Gilbert sequencing of end-labelled genomic DNA reveals a $(TTAGGG)_n$ ladder for both species (Forney *et al.* 1987; Henderson and Blackburn 1989; Johansen *et al.* 1992). The sequencing gel in fact shows two TTAGGG ladders superimposed, one base-pair out of register with each other. The interpretation of this is that the telomeres do not terminate at random at any of the six bases of the repeat unit but at two preferred positions, which one could speculate might reflect telomerase 'pausing' (see Chapter 4). Most of the $(TTAGGG)_n$ is internal in *Physarum* and is insensitive to BAL31 (Forney *et al.* 1987).

Dictyostelium discoideum

The ends of the linear extrachromosomal rDNA molecules have been cloned and sequenced. They show what appears to be a largely random mixture of repeats that conform to an AG_{1-8} consensus (Emery and Weiner 1981).

Ascaris lumbricoides

Telomeres have been cloned from the somatic tissue of this nematode where they are formed *de novo* following chromatin diminution (see Chapter 6). The terminal repeat sequence is $(TTAGGC)_n$ and all the corresponding signal is BAL31-sensitive (Müller *et al.* 1991). This repeat also cross-hybridizes to BAL31-sensitive sequences in *C. elegans* (cited in Cangiano and La Volpe 1993). $(TTAGGC)_n$ has also been cloned from the human filarial parasite *Loa loa*, but in the absence of BAL31 analysis it remains unknown whether it is telomeric (Klion *et al.* 1991).

Parascaris univalens

This is another nematode that undergoes chromatin diminution. Relative to the germ-line, somatic cells have eliminated around 85% of the genome, losing almost all the highly repetitive satellite DNA. Cloned germ-line satellite is composed of tandem repeats of TTGCA (Teschke *et al.* 1991). This sequence is a good candidate for the terminal repeat sequence of the somatic cells, as it is present in a plasmid end-library and detects a BAL31-sensitive smeared signal.

If the terminal repeat sequence is indeed TTGCA, it would be a most unusual repeat, having no strand bias for guanine and no runs of guanine.

Although it was isolated from an end-library such an approach only enriches for telomeric sequences and does not prevent internal sequences being cloned. It therefore remains a formal possibility that the somatic chromosomes terminate in another sequence that is distal to the $(TTGCA)_n$ repeats, as the clones isolated are much smaller than most of the BAL31-sensitive fragments. However, this is often the case with terminal repeat sequences because of the instability of large arrays in bacteria, and such reservations regarding the sequence at the very end of the chromosome apply to many species.

Insects

$(TTAGG)_n$ has been isolated from the silkworm *Bombyx mori* by screening a library with a $(TTNGGG)_5$ probe. Most of the signal in the silkworm genome is a BAL31-sensitive smear (Okazaki *et al.* 1993). A $(TTAGG)_5$ probe detects all the silkworm telomeres using *in situ* hybridization. It also detects signal in eight out of 11 orders of insects, but detects nothing in any dipteran including *Drosophila melanogaster*.

Drosophila melanogaster

No simple tandem repeat sequence has been found in this species despite numerous attempts to find a BAL31-sensitive signal by cross-hybridization to repeat sequences from other species (see for example Allshire *et al.* 1988; Richards and Ausubel 1988; Meyne *et al.* 1989). This includes the $(TTAGG)_n$ repeats found at other insect telomeres which do not detect any signal in *Drosophila*. Instead telomere sequence addition may be provided by large transposon-like elements (Chapter 9).

Chironomus

No short terminal repeat sequence has been identified in this dipteran. In *C. pallidivittatus* tandem repeat arrays of an underlying 340 bp sequence, corresponding to around 300 kb of sequence per telomere, are found by *in situ* hybridization at all except one telomere (Saiga and Edström 1985; Cohn and Edström 1991). A similar repeat is found in *C. tentans*, although it is 10 bp longer. It has been suggested that this 350 bp sequence evolved from a 165 bp sequence by acquiring a transposed element. One feature of this 165 bp sequence is a marked asymmetry in the strand distribution of guanine and it appears to be composed of degenerate short repeats. As short repeats with a strand asymmetry for guanine are a common feature of telomeric repeat sequences, Nielsen and Edström (1993) speculate that this 165 bp sequence may have evolved from an ancestral terminal repeat sequence such as the $(TTAGG)_n$ repeats still found in other insects.

In *C. thummi* there is a 176 bp sequence present in large tandem arrays at all except one telomere, although there is little sequence similarity

between it and the 340 bp *C. pallidivittatus* repeat (Carmona *et al.* 1985).
It has not yet been possible to determine if *Chironomus* chromosomes
terminate in any of these sequences as BAL31 analysis is not feasible for
such large arrays. It therefore remains to be seen if any of these repeats
are simply subtelomeric or whether they contribute to telomere function in
a fashion analogous to the situation in *Drosophila*, another dipteran.

Vertebrates

The structure of human and mouse telomeres is detailed in Chapter 10. The
$(TTAGGG)_n$ repeats found at human telomeres have been found at the
telomeres of every vertebrate so far tested by *in situ* hybridization (Wurster-
Hill *et al.* 1988; Meyne *et al.* 1989; Luke and Verma 1993). This is insufficient to prove that all vertebrates have $(TTAGGG)_n$ as their terminal
repeat sequence, as discussed in the opening section of this chapter using
the example of whale and cat satellite sequences.

Plants

Arabidopsis thaliana

The telomeres of this dicotyledonous plant were the first to be cloned from
a higher eukaryote, and illustrate the problems of isolating telomeres in
such species (Richards and Ausubel 1988). In lower eukaryotes, with a large
number of telomeres relative to the size of the genome, constructing a
plasmid 'end-library' has often been successful. In such an approach a
plasmid vector is blunt-end ligated to intact high molecular weight genomic
DNA. This is then digested with a restriction enzyme to release fragments
which can subsequently be circularized and transformed into bacteria. Only
terminal restriction fragments should in theory be ligated to the plasmid in
the first step and thus be cloned. However, in practice such libraries are
only enriched for telomeric clones. For example, only a small fraction of
the clones from a *T. brucei* end-library were telomeric (Blackburn and
Challoner 1984). A reasonable prediction is that the efficiency of enrich-
ment for telomeric sequences will decrease as the ratio of genome size
to the number of telomeres increases. For example, *T. brucei* has 200
telomeres for a haploid genome size of about 40 Mb, whereas a human has
46 for a genome of 3000 Mb.

Richards and Ausubel (1988) reasoned that telomeric sequences would be
repeated and enriched their primary end-libraries for such sequences. They
did this by producing two independent end-libraries in bacteriophage M13.
They then hybridized single-stranded DNA from the two libraries together
and selected annealed molecules. The inserts of 300 clones that came
through this process were screened by using them as hybridization probes

on Southern blots, with only one detecting a BAL31-sensitive smear characteristic of a telomeric signal. When sequenced, this clone contained a repeat array of the sequence $(TTTAGGG)_n$.

Although successful in this example an end-library approach produces small fragments and is technically laborious. It is not therefore suitable for the isolation of a large number of telomeric clones. To circumvent these problems Richards *et al.* (1992) have used a 1/2-YAC telomere cloning approach to isolate *Arabidopsis* telomeres in yeast, enabling much larger clones to be isolated. In addition to the terminal repeat arrays these clones also contain subtelomeric sequences, and partial sequence data from two clones have been obtained. Interestingly, one clone shows a direct transition from $(TTTAGGG)_n$ repeats to unique sequence with no evidence of any subtelomeric repetitive sequence elements. The other telomere possesses subtelomerically repeated sequences that are present at a number of different *Arabidopsis* telomeres. These subtelomeric sequences are highly methylated in the genome. However, even at this telomere single-copy sequence is reached within 2 kb of the terminal repeat array. It therefore appears that there are not extensive amounts of subtelomeric repetitive sequences in *Arabidopsis*.

Allium

The terminal repeat arrays have not been cloned but are likely to be similar to those in other plants. Tandem repeat arrays of an underlying 375 bp repeat unit form heterochromatic domains at all except two telomeres as judged by *in situ* hybridization and comprise over 4% of the entire genome (Barnes *et al.* 1985). This sequence may well be subtelomeric but it is not clear how close it is to the terminus because of the limited resolution of *in situ* hybridization.

Secale cereale

The *Arabidopsis* $(TTTAGGG)_n$ terminal repeats hybridize to *Secale cereale* (rye) telomeres *in situ* (Schwarzacher and Heslop-Harrison 1991). There are also large telomeric heterochromatin blocks which account for over 10% of the genome (Bedbrook *et al.* 1980). These are composed of a variety of short repeats, less then 1 kb in length, arranged as tandem repeat arrays.

Tomato

The terminal repeat sequence in tomato has been sequenced directly from genomic DNA (see above). It is TT(T/A)AGGG, similar to the *Arabidopsis* consensus sequence of TTTAGGG (Ganal *et al.* 1991), and almost all the signal is BAL31-sensitive. It is not possible to deduce from genomic sequencing whether the two variants are interspersed with each other at random. The terminal repeat arrays are both large (20–50 kb) and genetically hypervariable, even in self-pollinated highly inbred plants, a situation reminiscent of mouse telomeres (see Chapter 10). Melon also shows a similar telomere repeat array hypervariability. New size alleles arise frequently (up to 2% per generation) for arrays of both the terminal repeats and a 162 bp subtelomeric sequence. Arrays of 500 to 10 000 copies of this subtelomeric sequence are found at 20 of the 24 tomato telomeres and correspond to the heterochromatic terminal knobs of pachytene chromosomes. This 162 bp sequence is found only in tomato and closely related species. Because the tomato shows relatively few strain-specific polymorphisms, variability of the terminal and subterminal repeat arrays has been useful for obtaining restriction fragment length polymorphisms (RFLPs), and a number of telomeres have now been placed on the tomato genetic map (Broun *et al.* 1992; Ganal *et al.* 1992).

Maize and pea

The *Arabidopsis* (TTTAGGG)$_n$ terminal repeats detect a BAL31-sensitive signal in maize (Richards and Ausubel 1988; Burr *et al.* 1992). (TTTAGGG)$_n$ has therefore been found in both monocotyledonous and dicotyledonous plants. (TTTAGGG)$_n$-adjacent subtelomeric sequences have been isolated by a polymerase chain reaction (PCR)-based strategy (Burr *et al.* 1992). There are a few hundred copies of each sequence in the maize genome and they have been used to identify and map RFLPs using recombinant inbred strains. Eleven of the 13 mapped RFLPs are linked to markers at the end of eight of the 20 chromosome arms. Similarly, Ellis *et al.* (1992) have recently mapped at least one polymorphic marker detected with a (TTTAGGG)$_n$ probe to the end of a pea linkage group using recombinant inbred strains.

Barley

(TTTAGGG)$_n$ detect the telomeres and no interstitial sites in barley by *in situ* hybridization (Schwarzacher and Heslop-Harrison 1991). A 118 bp sequence has been isolated that forms large repeat arrays at 13 of the 14 telomeres by *in situ* hybridization. Some of this sequence is sufficiently close (<1 Mb) to the terminal repeat arrays that they can be present on the same restriction fragment using PFGE. The size of these fragments is polymorphic and RFLPs detected with either the 118 bp sequence or a (TTTAGGG)$_n$ probe have been used to place seven of the barley telomeres on the genetic map (Röder *et al.* 1993).

Chlamydomonas reinhardtii

The telomeres of this unicellular alga have been isolated from a plasmid end-library. The terminal repeat arrays, estimated to be about 300 bp in length, consist of (TTTTAGGG)$_n$ repeats and detect a BAL31-sensitive signal. Small amounts of subtelomeric sequence are also present on these end-clones and detect a subset of chromosomes on pulsed field gels (Petracek *et al.* 1990).

Biophysical properties of telomeric DNA

The guanine tetrad

Oligonucleotides corresponding to many terminal repeat sequence G-strands show unusual *in vitro* properties in that they can fold or associate to form both intramolecular and intermolecular structures stabilized by G·G base-pairing. It is possible that some of these structures could form *in vivo* and have a role in telomere function.

Since the turn of the century guanosine and derivatives such as 5'-GMP have been known to behave strangely *in vitro* in that they readily form viscous gels. X-ray fibre diffraction studies of these gels over 30 years ago (reviewed by Guschlbauer *et al.* 1990) revealed molecules arranged in what is now termed a G-tetrad structure (Fig. 3.2). G-tetrads are stabilized by Hoogsteen G·G pairing; this is a different mode of pairing from that found for guanine in the more conventional Watson–Crick G·C pairing (Fig. 3.3). Four atoms (H^1, N^7, O^6, and one of the exocyclic amino protons) are essential for Hoogsteen G·G pairing. The ability to form such a planar tetrad appears to be unique to guanosine and its derivatives, reflecting a fortuitous arrangement of multiple hydrogen bonding donor and acceptor sites. Poly(dG) also behaves in an unusual fashion, forming a very tight four-stranded structure stabilized by G-tetrads. In a similar fashion stretches of poly(dG) in denatured plasmids can fold into four-

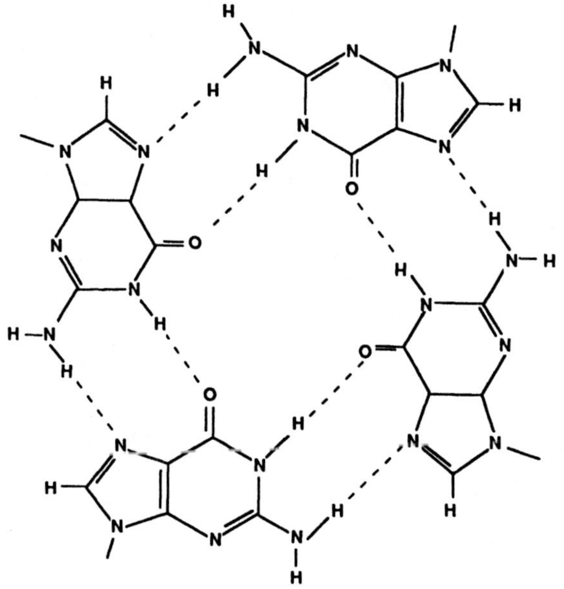

Fig. 3.2 The G-tetrad.

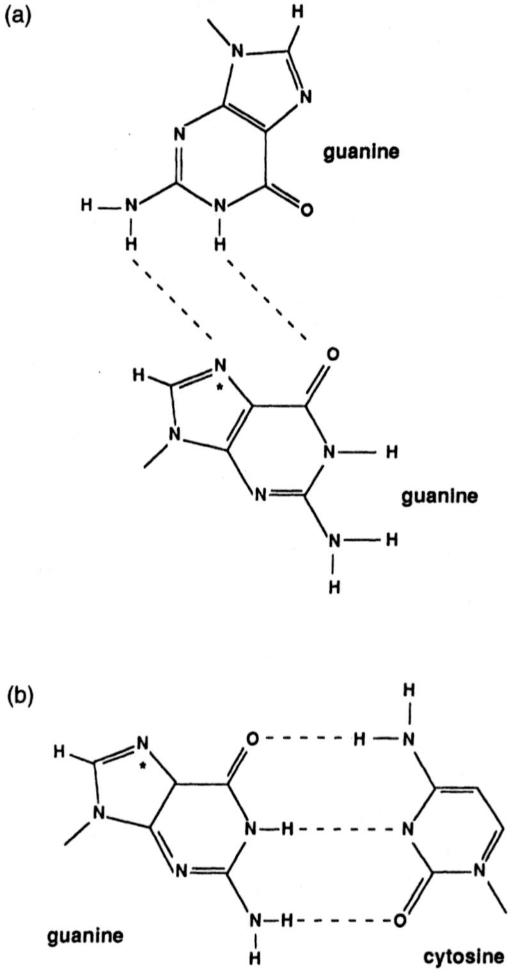

(a)

(b)

Fig. 3.3 Hydrogen bonding modes for guanine. Guanine–guanine Hoogsteen pairing is illustrated in (a). Guanine–cytosine Watson–Crick pairing is shown in (b). The N^7 atom important for Hoogsteen pairing is asterisked.

stranded antiparallel structures (Fig. 3.4) stabilized by G·G bonds (Panyutin *et al.* 1989, 1990).

Gels and polymers containing G-tetrads are well known to be stabilized by certain cations, with K^+ the best of the monovalent and Sr^{2+} the best of the divalent; both have very similar ionic radii. The suggested model is a crown-ether-like structure with a K^+ or Sr^{2+} cation sandwiched between two G-tetrads; Na^+ is too large to lodge.

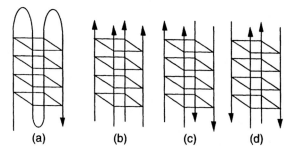

Fig. 3.4 Potential four-stranded structures. (a) Fold-back G-quartet. (b) Parallel quadruplex. (c) and (d) Antiparallel quadruplexes. Reprinted with permission from *Nature* (Aboul-ela *et al.* 1992), © 1992 Macmillan Magazines Ltd.

One of the first clues that a G-rich oligonucleotide is forming a secondary structure stabilized by G-tetrads is often that it runs with an aberrant mobility on a non-denaturing polyacrylamide gel. A number of biochemical methods can then be used to study its structure. For example, N^7 guanine methylation interferes with Hoogsteen G·G pairing but not conventional Watson–Crick G·C pairing (see Fig. 3.3). Therefore the sensitivity of a structure to N^7 methylation by dimethyl-sulphate (DMS) implies the existence of G·G pairing. It is also possible to perform methylation-protection assays in an analogous fashion, because when the N^7 of guanine is involved in Hoogsteen G·G pairing it is not accessible to methylation. Similarly, removal of the 2-amino group by replacement of guanine by its relative inosine preferentially destabilizes Hoogsteen pairing. Thus if a structure is sensitive to inosine replacement it suggests that G·G pairing is involved. Stabilization of structures by K^+ but not Na^+ may suggest the presence of stacked G-tetrads. Finally a range of spectroscopic methods can produce detailed information regarding the conformation of these structures.

Intramolecular fold-back structures

Complexes that migrated more quickly than the single-stranded molecule were detected when a range of oligonucleotides corresponding to a variety of terminal repeat G-strands were run on non-denaturing gels (Henderson *et al.* 1987). The formation of these more rapidly migrating complexes was independent of concentration. This first-order kinetics suggest some form of fold-back intramolecular structure. Such a fold-back G-quartet is shown schematically in Fig. 3.4a. Other evidence also pointed to a model where a single molecule was folding back on itself to form a four-stranded structure containing G-tetrads. This included NMR data indicating the presence of guanines in both *syn* and *anti* conformations, and UV crosslinks between thymines in different repeats of the same molecule (in Fig. 3.5 these

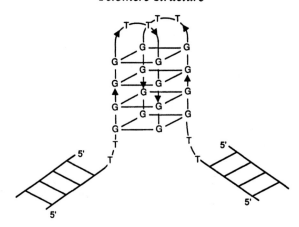

Fig. 3.5 Dimerization by quadruplex formation between two hairpin loops. Reprinted with permission from *Nature* (Sundquist and Klug 1989), © 1989 Macmillan Magazines Ltd.

thymines are adjacent to each other at the top of the loop). Furthermore, the more rapidly migrating complexes of both $(TTGGGG)_4$ and $(TTTTGGGG)_4$ form in the presence of Na^+, K^+ and Cs^+ but not Li^+ (Williamson *et al.* 1989). This is consistent with a model where the cation is sandwiched between two stacked G-tetrads, the cation forming electrostatic bonds with the keto groups of the four guanines both above and below it. Atomic sizes are such that the central pocket between the two G-tetrads can accommodate Na^+, K^+, and Cs^+ cations and allow them to form bonds with the guanines. However, although Li^+ does fit it is too small to be able to form stable bonds with all the guanines simultaneously. These steric constraints explain why Li^+ does not stabilize this fold-back structure. Similar fold-back intramolecular G-quartet structures can also be formed by oligonucleotides corresponding to the human $((TTAGGG)_4)$, *Chlamydomonas* $((TTTTAGGG)_4)$, and *Arabidopsis* $((TTTAGGG)_4)$ terminal repeat sequences (Balagurumoorthy *et al.* 1992; Petracek and Berman 1992).

Intermolecular structures

Many regions of the human genome, such as some promoters or the immunoglobulin switch region, contain G-rich sequences. Sen and Gilbert (1988) have studied oligonucleotides corresponding to regions of the immunoglobulin switch regions containing the motif d(GGGGAGCTGGGG). These oligonucleotides can form secondary structures that migrate *more slowly* on non-denaturing gels than when single-stranded. The guanines are protected from N^7 methylation by DMS when formed into such structures, suggesting Hoogsteen G·G pairing. The model proposed, G4-DNA, contains four parallel strands of DNA held together by G-tetrads and is illustrated in Fig. 3.4b.

One of the earliest suggestions that telomeric sequences could form intermolecular four-stranded structures came from the observation that purified *Oxytricha* and *Stylonychia* macronuclear DNA aggregated at high concentration in an Na^+-dependent manner (Lipps 1980; Lipps *et al.* 1982). This could not be straightforward Watson–Crick pairing because it was known that the 3' overhangs were not complementary, so a four-stranded association was suggested. This model invoked both G and C strands of the telomeres in the four-stranded structure (Lipps 1980; Lipps *et al.* 1982), whereas it is now believed that they involve only the two G-strands. Nevertheless, this work first raised the possibility of four-stranded structures at telomeres.

Oligonucleotides terminating in *Tetrahymena* (TTGGGG)$_2$ or *Oxytricha* (TTTTGGGG)$_2$ single-stranded 3' overhangs will form into a variety of dimerization or tetramerization products that are believed to be in dynamic equilibrium with each other (Sundquist and Klug 1989; Sen and Gilbert 1990; see also Hardin *et al.* 1991). The formation of these intermolecular structures appears to be promoted in some cases by Na^+ as these more slowly migrating complexes form in a Na^+-dependent fashion. However, once formed these structures are strongly stabilized against thermal denaturation by K^+. It is suggested that K^+ stabilizes the intra-strand structures to the detriment of the formation of inter-strand complexes. Some of the structures proposed, including parallel four-stranded structures and a dimer formed by association between two hairpin loops, are shown in Figs 3.4 and 3.5.

Intermolecular structures stabilized by G-tetrads could have a number of roles *in vivo*. For example, a parallel guanine quadruplex association could initiate the alignment of four sister chromatids at meiosis, even though the cytological evidence points to pairing initiating in subterminal regions rather than the terminal repeat arrays (see Chapter 2). These associations have significance to understanding the behaviour of purified *Oxytricha* genomic DNA. The macronuclear genome is composed of short molecules, and *in vitro* these slowly cohere at high concentrations to form multimeric complexes. This requires the presence of Na^+, and once formed these structures are strongly stabilized by K^+ (Oka and Thomas, 1987). The macronuclear telomeres terminate in (TTTTGGGG)$_2$ single-stranded overhangs and this is all that appears to be required for coherence, as it can be duplicated *in vitro* using oligonucleotides (Acevedo *et al.* 1991).

Oligonucleotides that form four-stranded structures can also form higher-order structures under the appropriate conditions. For example, whereas T_8G_3T forms a unique G4-DNA product, T_9G_3 generates a ladder of additional products containing eight, 12, and 16 oligonucleotide strands. It is suggested that such forms consist of tetramers assembling into higher-order structures with the G4-DNA regions interacting in a head-to-tail fashion (Sen and Gilbert 1992; see also Lu *et al.* 1992). Because of this an important requirement to form such superstructures is that the oligonucleotide terminates in a G residue, and thus T_9G_3 but not T_8G_3T forms these higher-order structures. These may also be relevant to the association of *Oxytricha* genomic DNA *in vitro*.

Recent advances: from models to crystals

Detailed structural and chemical analyses of both intramolecular and inter-molecular structures over the last few years have increased our biochemical understanding of these structures (Guo *et al.* 1992*a,b*; Scaria *et al.* 1992; Lu *et al.* 1993). Many early models were based on a combination of the known G-tetrad structure and the limited biochemical data that were available. There has now been extensive revision of these models following elucidation of the X-ray structure of $G_4T_4G_4$ in fold-back G-quartet form (Kang *et al.* 1992), detailed NMR data of $G_4T_4G_4$ and $G_4T_4G_4T_4G_4T_4G_4$ in fold-back G-quartet form (Smith and Feigon 1992), and NMR solution structures of T_4G_4 in parallel tetramer (G4-DNA) form (Gupta *et al.* 1993; Wang and Patel 1993). These data indicated a distribution of *syn* and *anti* glycosidic angles, alternating along each strand, that was different from the original models. In contrast to these longer oligonucleotides, NMR studies of short oligonucleotides such as TTAGGG, TTGGGG, and TGGGGT in solution indicate that they assemble into a parallel-stranded G-quadruplex structure with all the guanines in *anti* conformation, as is found in more conventional B-DNA (Aboul-ela *et al.* 1992; Wang and Patel 1992). It is suggested that the presence of *syn* guanines in the fold-back structures reflects steric constraints introduced within antiparallel strands.

Other unusual structures

Double-stranded $(TTGGGG)_n$ cloned into a plasmid vector forms an atypical DNA conformation when under negative superhelical stress, becoming sensitive to nuclease S_1 (Budarf and Blackburn 1987). The unusual structure forms at low pH (< 4.5) and chemical modifications suggest that the C-strand forms a hairpin predominantly stabilized by protonated $C \cdot C^+$ base pairs (hence the pH requirement). In contrast the G-strand appears to remain single-stranded (Lyamichev *et al.* 1989). Such open regions presumably explain their S_1 nuclease sensitivity. Similarly, $(CCCCAA)_4$ and $(CCCCAAAA)_4$ oligonucleotides can form snap-back structures *in vitro* using $C \cdot C^+$ pairing (Ahmed and Henderson 1992).

A very bizarre structure is formed by the short oligonucleotide TCCCC at acid pH. High-resolution NMR of this four-stranded structure indicates that it consists of two base-paired parallel duplexes closely associated with each other. The two duplexes are not base-paired with each other but instead intercalate and interdigitate with each other (Gehring *et al.* 1993). This structure has been called i-DNA; although it is possible that the C-strand associations described above involve this novel structure there is no evidence for this as yet.

In vivo significance

Are these G-tetrad-containing structures an *in vitro* oddity or do they have an *in vivo* role at telomeres? A number of approaches have been taken to ask whether such structures form *in vivo*, and if they do what their effect might be on telomere function.

Is formation of a G-tetrad structure required for telomerase function? It has been shown that *Oxytricha* telomerase is unable to use (TTTTGGGG)$_4$ as a primer when it is formed into a fold-back G-quartet structure (Zahler *et al.* 1991). Therefore formation of a fold-back G-quartet structure, rather than being necessary for a sequence to function as a telomerase primer, actually appears to inhibit the enzyme. *Tetrahymena* telomerase can use oligonucleotides with inosine substituted for guanine (Henderson *et al.* 1990). As these oligonucleotides cannot form G-tetrad structures this again argues that G-tetrad structures are not required in a telomerase primer. The *Oxytricha* telomere binding protein is unable to bind (TTTTGGGG)$_4$ folded into a fold-back G-quartet (Raghuraman and Cech 1990). The ability of sequences to be healed by telomere addition in *S. cerevisiae* does not correlate with their ability to form G-tetrad structures. Sequences which cannot form G·G base pairs still function as efficient substrates for TG$_{1-3}$ repeat addition (Lustig 1992).

It therefore appears that intramolecular G-tetrad structures are not required for any of the telomere-associated functions that have been tested both *in vitro* and *in vivo*. However, as they form in a concentration-independent fashion at salt concentrations not too different from what might be found in the cell, they could possibly form *in vivo*. If so they could have a detrimental effect on, for example, telomerase function. A single-stranded terminus that could form into a G-quartet could be the target for a specific nuclease, and indeed there is preliminary evidence for such an activity in yeast (see below).

G-tetrad structures may be detrimental to telomere function if they inhibit telomerase action or the binding of telomere proteins. Liu *et al.* (1993) provide preliminary evidence consistent with the existence of a yeast nuclease specific for G-tetrad structures which could remove any termini that have folded into such structures, releasing ends in the appropriate conformation to act as substrates for binding proteins and telomerase. In yeast the TG_{1-3} strand of at least some telomeres becomes single-stranded transiently following replication (see Chapter 4) and appears to result in telomere–telomere interactions such that linear plasmids migrate on gels as nicked circles at this point in the cell cycle (Wellinger *et al.* 1992, 1993*a,b*). Although there is no evidence to support this as yet, one could speculate that this nuclease functions to resolve such interactions.

In contrast to fold-back structures, intermolecular structures require very high DNA concentrations *in vitro* and for this reason might not be expected to form *in vivo*. For example, formation of the intermolecular structures that result in the coherence of *Oxytricha* telomeres requires high DNA concentrations and long incubation times. However, a number of basic proteins including histone H1 and the β subunit of the *Oxytricha* telomere-binding protein promote rapid intermolecular G-tetrad structure formation of *Oxytricha* telomeric oligonucleotides *in vitro* at much lower DNA concentrations (Fang and Cech 1993*a,b*; Sundquist 1993). The rate of assembly of the G-tetrad structure, usually a slow kinetic process, is increased by as much as a million-fold. The β subunit does not shift the equilibrium position but kinetically accelerates G-tetrad structure formation at nanomolar DNA concentrations such that it has a $t_\frac{1}{2}$ of about 1 h at physiological salt concentrations. Thus G-tetrad formation is now able to proceed at physiological salt concentrations with a more realistic DNA concentration and reaction rate. However, there is no direct evidence that *Oxytricha* telomeres are folded into G-tetrad structures *in vivo*, as judged by the *in vivo* DMS protection pattern (Price and Cech 1987), which is different from that formed by the corresponding oligonucleotides folded into a G-tetrad structure (Fang and Cech 1993*a*). Because the equilibrium position is not altered, the β subunit can in principle also accelerate the unfolding of an intermolecular G-tetrad structure which has formed. Such an unfolded form may then be stabilized by binding of the α subunit of the telomere binding protein, which is unable to bind folded structures (Raghuraman and Cech 1990). Thus the β subunit may facilitate the binding of the α subunit by accelerating the unfolding of any G-tetrad structures which may have formed. Removal of G-tetrad structures will also have the effect of ensuring that the chromosome end is not folded into a structure that would inhibit telomerase.

An evolutionary argument for an *in vivo* role for G-tetrad structures is the conservation of terminal repeat structures. Some but not all terminal repeat sequences have a G-strand which has the potential to form G-tetrad structures, and furthermore this strand is always oriented on the chromosome so that the primer gap is likely to leave it single-stranded, even if only for a short period immediately following replication. This could imply that the ability to form G-tetrad structures is an essential telomere repeat function and is therefore being selected for. However, it could instead reflect some biochemical feature of telomerase such that it functions optimally with a G-rich template (Blackburn, 1992). In this case the ability to form a G-tetrad structure would be a by-product of the sequence requirements for telomerase function and would not have an *in vivo* role.

Proteins that recognize structures containing G·G base pairs

The argument that G-tetrad structures can form *in vivo* is strengthened by the identification of proteins such as QUAD (Weisman-Shomer and Fry, 1993) that bind such structures. Although most of these proteins are not likely candidates for having a role at telomeres, their existence does suggest that G-tetrad structures can form and have a role *in vivo*, even if it is not at telomeres.

Another example of a protein that recognizes structures containing G·G base-pairs is MyoD, a vertebrate transcription factor involved in the initiation of myogenesis (Walsh and Gualberto 1992). There is a binding site for this protein in the mouse muscle-specific creatine kinase enhancer which is recognized when the DNA is single-stranded. However, an oligonucleotide corresponding to the binding site and which terminates in ... GTTGGGGGA-3′ forms a more slowly migrating (probably intermolecular) structure stabilized by guanine tetrads. It is this form that is predominantly bound by MyoD. Consistent with this is the observation that N^7 guanine methylation prevents MyoD binding. The affinity of MyoD for its recognition sequence is high (K_D of *c.* $10^{10} M^{-1}$). MyoD can bind molecules terminating in $(TTGGGG)_2$ when assembled into an intermolecular G-tetrad structure (Walsh and Gualberto 1992). The affinity of this latter reaction was not reported, and one is tempted to speculate that it is caused by MyoD recognizing its binding site when single-stranded and formed into a structure stabilized by G-tetrads; the binding reaction would then be analogous to the way that sequence-specific proteins that bind B-form DNA also show lower affinity, non-specific binding for B-DNA sequences in general. Although it is unlikely that a tissue-specific transcription factor has any role at telomeres, it does suggest that G-tetrad structures have some as yet unknown role in MyoD function at the creatine kinase enhancer.

An avian protein termed MF3 has been identified, using a band-shift assay, that binds a range of single-stranded oligonucleotides including TTAGGG repeats *in vitro* with high affinity ($K_D \approx 10^9 M^{-1}$) (Gualberto *et al.* 1992). The common feature among these oligonucleotides, which are recognized some four orders of magnitude better than unrelated oligonucleotides, is that they can form higher-order structures involving G·G bonds. N^7 guanine methylation or inosine substitution abolishes factor binding. These observations are consistent with the factor binding to structures stabilized by G·G base pairs. All these proteins support the argument that G-tetrad structures form and have a role *in vivo*, although this may not be at telomeres.

Summary

Eukaryotic chromosomes usually terminate in tandem repeats of a short sequence. These are implicated in essential telomere function by being found at all chromosome ends in a cell. A number of structural features are common to many terminal repeat sequences. In the next chapter the ability of models of telomere synthesis to explain these various features will be discussed.

1. Of those species whose telomeres have been successfully studied, the majority show a terminal array of short repeats of the form T_xG_y or T_xAG_y. The main variation comes in the number of Ts and Gs, such as the TG_{1-3} (*Saccharomyces*), TTTTGGGG (*Euplotes*), TTAGG (*Bombyx*), and TTTTAGGG (*Chlamydomonas*). Curiously there is never more than a single A.

2. The repeats show a strand asymmetry in base composition. The G-rich strand is oriented 5′ to 3′ towards the end of the chromosome for all species studied to date.

3. The terminus of the G-rich strand is single-stranded for at least some of the cell cycle, although it has not been possible to address this question in most species.

4. In a population of cells the size of repeat array at any one telomere can be very exact (e.g. in *Euplotes*) or vary from cell to cell (e.g. in humans).

5. There is a greater rate of sequence turnover in the region of a terminal repeat array which is closer to the end of the chromosome. This has been shown in *S. cerevisiae* and *Plasmodium* by cloning the same telomere repeatedly from a population of cells.

6. Some species have heterogeneous repeat sequences, such as *Saccharomyces* and *Dictyostelium*.

7. Some species have two repeat sequences interspersed with each other, such as *Plasmodium* and *Paramecium*.

In contrast to the terminal repeats, subtelomeric sequences are characterized by enormous variation in sequence between species. This species specificity, together with often being found at only a subset of telomeres, has made it difficult to ascribe an *in vivo* role to them.

It is important to consider the possibility that telomeres consisting of repeats of a short sequence are not as universal as this chapter might suggest. One wonders how many attempts to clone telomeric sequences in other species failed and are unrecorded because a simple repeat was being sought, whereas one did not in fact exist. In most species telomeric sequences have been isolated by their similarity to terminal repeat sequences in other species; most approaches have used either some variation on cross-hybridization or the ability of short TG-rich sequences to seed terminal repeat addition in yeast. However, it is now clear that *Drosophila* does not have a simple short repeat sequence, and cross-hybridization approaches have failed to isolate the telomeric sequences of this species (Chapter 9). Three out of 11 orders of insects do not have short terminal sequence as judged by cross-hybridization to the insect (TTAGG)$_n$ terminal sequence (Okazaki *et al.* 1993), and it is possible that there are a number of other species that do not have a simple terminal repeat sequence.

References

Reviews

Blackburn, E. H. (1992). Telomerases. *Annu. Rev. Biochem.*, **61**, 113–29.

Foote, S. J. and Kemp, D. J. (1989.) Chromosomes of malaria parasites. *Trends Genet.*, **5**, 337–42.

Guschlbauer, W., Chantot, J.-F., and Thiele, D. (1990). Four-stranded nucleic acid structures 25 years later: from guanosine gels to telomer DNA. *J. Biomol. Struc. Dynam.*, **8**, 491–511.

Sundquist, W. I. (1993). Conducting the G-quartet. *Curr. Biol.*, **3**, 893–5.

Primary papers

Aboul-ela, F., Murchie, A. I. H., and Lilley, D. M. J. (1992). NMR study of parallel-stranded tetraplex formation by the hexadeoxynucleotide d(TG$_4$T). *Nature*, **360**, 280–2.

Acevedo, O. L., Dickinson, L. A., Macke, T. J., and Thomas, C. A. (1991). The coherence of synthetic telomeres. *Nucleic Acids Res.*, **19**, 3409–19.

Adam, R. D., Nash, T. E., and Wellems, T. E. (1991). Telomeric location of *Giardia* rDNA genes. *Mol. Cell. Biol.*, **11**, 3326–30.

Adegoke, J. A., Árnason, Ú., and Widegren, B. (1993). Sequence organization and evolution, in all extant whalebone whales, of a DNA satellite with terminal chromosome localization. *Chromosoma*, **102**, 382–8.

Ahmed, S. and Henderson, E. (1992). Formation of novel hairpin structures by telomeric C-strand oligonucleotides. *Nucleic Acids Res.*, **20**, 507–11.

Allshire, R. C., Gosden, J. R., Cross, S. H., Cranston, G., Rout, D., Sugawara, N., *et al.* (1988). Telomeric repeat from *T. thermophila* cross hybridizes with human telomeres. *Nature*, **332**, 656–9.

Allshire, R. C., Dempster, M., and Hastie, N. D. (1989). Human telomeres contain at least three types of G-rich repeat distributed non-randomly. *Nucleic Acids Res.*, **17**, 4611–27.

Árnason, Ú. and Widegren, B. (1989). Composition and chromosomal localization of cetacean highly repetitive DNA with special reference to the blue whale, *Balaenoptera musculus. Chromosoma*, **98**, 323–9.

Árnason, Ú., Purdom, I. F., and Jones, K. W. (1978). Conservation and chromosomal localization of DNA satellites in Balenopterid whales. *Chromosoma*, **66**, 141–59.

Balagurumoorthy, P., Brahmachari, S. K., Mohanty, D., Bansal, M., and Sasisekharan, V. (1992). Hairpin and parallel quartet structures for telomeric sequences. *Nucleic Acids Res.*, **20**, 4061–7.

Barnes, S. R., James, A. M., and Jamieson, G. (1985). The organization, nucleotide sequence, and chromosomal distribution of a satellite DNA from *Allium cepa. Chromosoma*, **92**, 185–92.

Bedbrook, J. R., Jones, J., O'Dell, M., Thompson, R. D., and Flavell, R. B. (1980). A molecular description of telomeric heterochromatin in Secale species. *Cell*, **19**, 545–60.

Blackburn, E. H. and Challoner, P. B. (1984). Identification of a telomeric DNA sequence in Trypanosoma brucei. *Cell*, **36**, 447–57.

Blackburn, E. H. and Gall, J. G. (1978). A tandemly repeated sequence at the termini of the extrachromosomal ribosomal RNA genes in *Tetrahymena. J. Mol. Biol.*, **120**, 33–53.

Bourgain, F. M. and Katinka, M. D. (1991). Telomeres inhibit end to end fusion and enhance maintenance of linear DNA molecules injected into the *Paramecium primaurelia* macronucleus. *Nucleic Acids Res.*, **19**, 1541–7.

Broun, P., Ganal, M. W., and Tanksley, S. D. (1992). Telomeric arrays display high levels of heritable polymorphism among closely related plant varieties. *Proc. Natl. Acad. Sci. USA*, **89**, 1354–7.

Brown, W. R. A. (1989). Molecular cloning of human telomeres in yeast. *Nature*, **338**, 774–6.

Brown, W. R. A., MacKinnon, P. J., Villasanté, A., Spurr, N., Buckle, V. J., and Dobson, M. J. (1990). Structure and polymorphism of human telomere-associated DNA. *Cell*, **63**, 119–32.

Budarf, M. and Blackburn, E. (1987). S1 nuclease sensitivity of a double-stranded telomeric DNA sequence. *Nucleic Acids Res.*, **15**, 6273–92.

Burr, B., Burr, F. A., Matz, E. C., and Romero-Severson, J. (1992). Pinning down loose ends: mapping telomeres and factors affecting their length. *Plant Cell*, **4**, 953–60.

Button, L. L. and Astell, C. R. (1986). The *Saccharomyces cerevisiae* chromosome III left telomere has a type X, but not a type Y′, ARS region. *Mol. Cell. Biol.*, **6**, 1352–6.

Cangiano, G. and La Volpe, A. (1993). Repetitive DNA sequences located in the terminal portion of the *Caenorhabditis elegans* chromosomes. *Nucleic Acids Res.*, **21**, 1133–9.

Carmona, M. J., Morcillo, G., Galler, R., Martinez-Salas, E., de la Campa, A. G., Diez, J. L., and Edström, J. E. (1985). Cloning and molecular characterization of a telomeric sequence from a temperature-induced Balbiani ring. *Chromosoma*, **92**, 108–115.

Challoner, P. B. and Blackburn, E. H. (1986). Conservation of sequences adjacent to the telomeric C_4A_2 repeats of ciliate macronuclear ribosomal RNA gene molecules. *Nucleic Acids Res.*, **14**, 6299-311.

Chan, C. S. M. and Tye, B.-K. (1983*a*). A family of *Saccharomyces cerevisiae* repetitive autonomously replicating sequences that have very similar genomic environments. *J. Mol. Biol.*, **168**, 505-23.

Chan, C. S. M. and Tye, B.-K. (1983*b*). Organization of DNA sequences and replication origins at yeast telomeres. *Cell*, **33**, 563-73.

Cherry, J. M. and Blackburn, E. H. (1985). The internally located telomeric sequences in the germ-line chromosomes of Tetrahymena are at the ends of transposon-like elements. *Cell*, **43**, 747-58.

Cohn, M. and Edström, J. E. (1991). Evolutionary relations between subtypes of telomere-associated repeats in *Chironomus*. *J. Mol. Evol.*, **32**, 463-8.

Corcoran, L. M., Thompson, J. K., Walliker, D., and Kemp, D. J. (1988). Homologous recombination within subtelomeric repeat sequences generates chromosome size polymorphisms in P. falciparum. *Cell*, **53**, 807-13.

Cross, S. H., Allshire, R. C., McKay, S. J., McGill, N. I., and Cooke, H. J. (1989). Cloning of human telomeres by complementation in yeast. *Nature*, **338**, 771-4.

Dawson, D. and Herrick, G. (1984). Telomeric properties of C_4A_4-homologous sequences in micronuclear DNA of Oxytricha fallax. *Cell*, **36**, 171-7.

de Bruin, D., Lanzer, M., and Ravetch, J. V. (1992). Characterization of yeast artificial chromosomes from *Plasmodium falciparum* construction of a stable, representative library and cloning of telomeric DNA fragments. *Genomics*, **14**, 332-9.

de Lange, T., Kooter, J. M., Michels, P. A. M., and Borst, P. (1983). Telomere conversion in trypanosomes. *Nucleic Acids Res.*, **11**, 8149-65.

de Lange, T., Shiue, L., Myers, R. M., Cox, D. R., Naylor, S. L., Killery, A. M., and Varmus, H. E. (1990). Structure and variability of human chromosome ends. *Mol. Cell. Biol.*, **10**, 518-27.

Dore, E., Pace, T., Ponzi, M., Picci, L., and Frontali, C. (1990). Organization of subtelomeric repeats in *Plasmodium berghei*. *Mol. Cell. Biol.*, **10**, 2423-7.

Dunn, B., Szauter, P., Pardue, M. L., and Szostak, J. W. (1984). Transfer of yeast telomeres to linear plasmids by recombination. *Cell*, **39**, 191-201.

Edman, J. C. (1992). Isolation of telomerelike sequences from *Cryptococcus neoformans* and their use in high-efficiency transformation. *Mol. Cell. Biol.*, **12**, 2777-83.

Ellis, T. H. N., Turner, L., Hellens, R. P., Lee, D., Harker, C. L., Enard, C., *et al.* (1992). Linkage maps in pea. *Genetics*, **130**, 649-63.

Emery, H. S. and Weiner, A. M. (1981). An irregular satellite sequence is found at the termini of the linear extrachromosomal rDNA in Dictyostelium discoideum. *Cell*, **26**, 411-19.

Fang, G. and Cech, T. R. (1993*a*). The β subunit of Oxytricha telomere-binding protein promotes G-quartet formation of telomeric DNA. *Cell*, **74**, 875-85.

Fang, G. and Cech, T. R. (1993*b*). Characterization of a G-quartet formation reaction promoted by the β-subunit of the Oxytricha telomere-binding protein. *Biochemistry*, **32**, 11646-57.

Fanning, T. G. (1987). Origin and evolution of a major feline satellite DNA. *J. Mol. Biol.*, **197**, 627-34.

Fasman, G. D. (1976). *Handbook of biochemistry and molecular biology* (3rd ed). CRC Press, Boca Raton, FL.

Forney, J., Henderson, E. R., and Blackburn, E. H. (1987). Identification of the telomeric sequence of the acellular slime molds *Didymium iridis* and *Physarum polycephalum*. *Nucleic Acids Res.*, **15**, 9143-52.

Ganal, M. W., Lapitan, N. L. V., and Tanksley, S. D. (1991). Macrostructure of the tomato telomeres. *Plant Cell*, **3**, 87-94.

Ganal, M. W., Broun, P., and Tanksley, S. D. (1992). Genetic mapping of tandemly repeated telomeric DNA sequences in tomato (*Lycopersicon esculentum*). *Genomics*, **14**, 444-8.

Gehring, K., Leroy, J.-L., and Guéron, M. (1993). A tetrameric DNA structure with protonated cytosine·cytosine base pairs. *Nature*, **363**, 561-5.

Gualberto, A., Patrick, R. M., and Walsh, K. (1992). Nucleic acid specificity of a vertebrate telomere-binding protein: evidence for G-G base pair recognition at the core-binding site. *Genes Dev.*, **6**, 815-24.

Guo, Q., Lu, M., and Kallenbach, N. R. (1992*a*). Adenine affects the structure and stability of telomeric sequences. *J. Biol. Chem.*, **267**, 15293-300.

Guo, Q., Lu, M., Marky, L. A., and Kallenbach, N. R. (1992*b*). Interaction of the dye ethidium bromide with DNA containing guanine repeats. *Biochemistry*, **31**, 2451-5.

Gupta, G., Garcia, A. E., Guo, Q., Lu, M., and Kallenbach, N. R. (1993). Structure of a parallel-stranded tetramer of the *Oxytricha* telomeric DNA sequence dT_4G_4. *Biochemistry*, **32**, 7098-103.

Hardin, C. C., Henderson, E., Watson, T., and Prosser, J. K. (1991). Monovalent cation induced structural transitions in telomeric DNAs: G-DNA folding intermediates. *Biochemistry*, **30**, 4460-72.

Henderson, E. R. and Blackburn, E. H. (1989). An overhanging 3′ terminus is a conserved feature of telomeres. *Mol. Cell. Biol.*, **9**, 345-8.

Henderson, E., Hardin, C. C., Walk, S. K., Tinoco, I., and Blackburn, E. H. (1987). Telomeric DNA oligonucleotides form novel intramolecular structures containing guanine-guanine base pairs. *Cell*, **51**, 899-908.

Henderson, E. R., Moore, M., and Malcolm, B. A. (1990). Telomere G-strand structure and function analyzed by chemical protection, base analogue substitution, and utilization by telomerase in vitro. *Biochemistry*, **29**, 732-7.

Herrick, G., Cartinhour, S., Dawson, D., Ang, D., Sheets, R., Lee, A., and Williams, K. (1985). Mobile elements bounded by C_4A_4 telomeric repeats in Oxytricha fallax. *Cell*, **43**, 759-68.

Horowitz, H. and Haber, J. E. (1984). Subtelomeric regions of yeast chromosomes contain a 36 base-pair tandemly repeated sequence. *Nucleic Acids Res.*, **12**, 7105-21.

Horowitz, H. and Haber, J. E. (1985). Identification of autonomously replicating circular subtelomeric Y′ elements in *Saccharomyces cerevisiae*. *Mol. Cell. Biol.*, **5**, 2369-80.

Horowitz, H., Thorburn, P., and Haber, J. E. (1984). Rearrangements of highly polymorphic regions near telomeres of *Saccharomyces cerevisiae*. *Mol. Cell. Biol.*, **4**, 2509-17.

Hunter, D. J., Williams, K., Cartinhour, S., and Herrick, G. (1989). Precise excision of telomere-bearing transposons during *Oxytricha fallax* macronuclear development. *Genes Dev.*, **3**, 2101-12.

Jäger, D. and Philippsen, P. (1989*a*). Many yeast chromosomes lack the telomere-specific Y′ sequence. *Mol. Cell. Biol.*, **9**, 5754-7.

Jäger, D. and Philippsen, P. (1989*b*). Stabilization of dicentric chromosomes in *Saccharomyces cerevisiae* by telomere addition to broken ends or by centromere deletion. *EMBO J.*, **8**, 247-54.

Javerzat, J.-P., Bhattacherjee, V., and Barreau, C. (1993). Isolation of telomeric DNA from the filamentous fungus *Podospora anserina* and construction of a self-replicating linear plasmid showing high transformation frequency. *Nucleic Acids Res.*, 21, 497–504.

Johansen, S., Johansen, T., and Haugli, F. (1992). Extrachromosomal ribosomal DNA of *Didymium iridis*: sequence analysis of the large subunit ribosomal RNA gene and sub-telomeric region. *Curr. Genet.*, 22, 305–12.

Johnson, E. M. (1980). A family of inverted repeat sequences and specific single-strand gaps at the termini of the Physarum rDNA palindrome. *Cell*, 22, 875–86.

Kämper, J., Meinhardt, F., Gunge, N., and Esser, K. (1989). In vivo construction of linear vectors based on killer plasmids from *Kluyveromyces lactis*: selection of a nuclear gene results in attachment of telomeres. *Mol. Cell. Biol.*, 9, 3931–7.

Kang, C., Zhang, X., Ratliff, R., Moyzis, R., and Rich, A. (1992). Crystal structure of four-stranded *Oxytricha* telomeric DNA. *Nature*, 356, 126–31.

Katinka, M. D. and Bourgain, F. M. (1992). Interstitial telomeres are hotspots for illegitimate recombination with DNA molecules injected into the macronucleus of *Paramecium primaurelia*. *EMBO J.*, 11, 725–32.

Katzen, A. L., Cann, G. M., and Blackburn, E. H. (1981). Sequence-specific fragmentation of macronuclear DNA in a holotrichous ciliate. *Cell*, 24, 313–20.

Klion, A. D., Raghavan, N., Brindley, P. J., and Nutman, T. B. (1991) Cloning and characterization of a species-specific repetitive DNA sequence from *Loa loa*. *Mol. Biochem. Parasitol.*, 45, 297–306.

Klobutcher, L. A., Swanton, M. T., Donini, P., and Prescott, D. M. (1981). All gene-sized DNA molecules in four species of hypotrichs have the same terminal sequence and an unusual 3′ terminus. *Proc. Natl. Acad. Sci. USA*, 78, 3015–19.

le Blancq, S. M., Kase, R. S., and van der Ploeg, L. H. T. (1991). Analysis of a *Giardia lamblia* rRNA encoding telomere with (TAGGG)$_n$ as the telomere repeat. *Nucleic Acids Res.*, 19, 5790.

Lipps, H. J. (1980). *In vitro* aggregation of the gene-sized DNA molecules of the ciliate *Stylonychia mytilus*. *Proc. Natl. Acad. Sci. USA*, 77, 4104–7.

Lipps, H. J., Gruissem, W., and Prescott, D. M. (1982). Higher order DNA structure in macronuclear chromatin of the hypotrichous ciliate *Oxytricha nova*. *Proc. Natl. Acad. Sci. USA*, 79, 2495–9.

Liu, Z., Frantz, J. D., Gilbert, W., and Tye, B.-K. (1993). Identification and characterization of a nuclease activity specific for G4 tetrastrand DNA. *Proc. Natl. Acad. Sci. USA*, 90, 3157–61.

Louis, E. J. and Haber, J. E. (1990a). The subtelomeric Y′ repeat family in *Saccharomyces cerevisiae*: an experimental system for repeated sequence evolution. *Genetics*, 124, 533–45.

Louis, E. J. and Haber, J. E. (1990b). Mitotic recombination among subtelomeric Y′ repeats in *Saccharomyces cerevisiae*. *Genetics*, 124, 547–59.

Louis, E. J. and Haber, J. E. (1992). The structure and evolution of subtelomeric Y′ repeats in *Saccharomyces cerevisiae*. *Genetics*, 131, 559–74.

Lu, M., Guo, Q., and Kallenbach, N. R. (1992). Structure and stability of sodium and potassium complexes of dT_4G_4 and dT_4G_4T. *Biochemistry*, 31, 2455–9.

Lu, M., Guo, Q., and Kallenbach, N. R. (1993). Thermodynamics of G-tetraplex formation by telomeric DNAs. *Biochemistry*, 32, 598–601.

Luke, S. and Verma, R. S. (1993). Telomeric repeat [TTAGGG]$_n$ sequences of human chromosome are conserved in chimpanzee (*Pan troglodytes*). *Mol. Gen. Genet.*, **237**, 460–62.

Lustig, A. J. (1992). Hoogsteen G-G base pairing is dispensable for telomere healing in yeast. *Nucleic Acids Res.*, **20**, 3021–8.

Lyamichev, V. I., Mirkin, S. M., Danilevskaya, O. N., Voloshin, O. N., Balatskaya, S. V, Dobrynin, V. N., *et al.* (1989). An unusual DNA structure detected in a telomeric sequence under superhelical stress and at low pH. *Nature*, **339**, 634–7.

McEachern, M. J. and Hicks, J. B. (1993). Unusually large telomeric repeats in the yeast *Candida albicans*. *Mol. Cell. Biol.*, **13**, 551–60.

Matsumoto, T., Fukui, K., Niwa, O., Sugawara, N., Szostak, J. W., and Yanagida, M. (1987). Identification of healed terminal DNA fragments in linear mini-chromosomes of *Schizosaccharomyces pombe*. *Mol. Cell. Biol.*, **7**, 4424–30.

Meyne, J., Ratliff, R. L., and Moyzis, R. K. (1989). Conservation of the human telomere sequence (TTAGGG)$_n$ among vertebrates. *Proc. Natl. Acad. Sci. USA*, **86**, 7049–53.

Meyne, J., Baker, R. J., Hobart, H. H., Hsu, T. C., Ryder, O. A., Ward, O.G., *et al.* (1990). Distribution of non-telomeric sites of the (TTAGGG)$_n$ telomeric sequence in vertebrate chromosomes. *Chromosoma*, **99**, 3–10.

Moyzis, R. K., Buckingham, J. M., Cram, L. S, Dani, M., Deaven, L. L., Jones, M.D., *et al.* (1988). A highly conserved repetitive DNA sequence, (TTAGGG)$_n$, present at the telomeres of human chromosomes. *Proc. Natl. Acad. Sci. USA*, **85**, 6622–6.

Müller, F., Wicky, C., Spicher, A., and Tobler, H. (1991). New telomere formation after developmentally regulated chromosomal breakage during the process of chromatin diminution in Ascaris lumbricoides. *Cell*, **67**, 815–22.

Murray, A. W. and Szostak, J. W. (1986). Construction and behavior of circularly permuted and telocentric chromosomes in *Saccharomyces cerevisiae*. *Mol. Cell. Biol.*, **6**, 3166–72.

Nielsen, L. and Edström, J.-E. (1993). Complex telomere-associated repeat units in members of the genus *Chironomus* evolve from sequences similar to simple telomeric repeats. *Mol. Cell. Biol.*, **13**, 1583–9.

Oka, Y. and Thomas, C.A. (1987). The cohering telomeres of Oxytricha. *Nucleic Acids Res.*, **15**, 8877–98.

Oka, Y., Shiota, S., Nakai, S., Nishida, Y., and Okubo, S. (1980). Inverted terminal repeat sequence in the macronuclear DNA of *Stylonychia pustulata*. *Gene*, **10**, 301–6.

Okazaki, S., Tsuchida, K., Maekawa, H., Ishikawa, H., and Fujiwara, H. (1993). Identification of a pentanucleotide telomeric sequence, (TTAGG)$_n$, in the silkworm *Bombyx mori* and in other insects. *Mol. Cell. Biol.*, **13**, 1424–32.

Pace, T., Ponzi, M., Dore, E., Janse, C., Mons, B., and Frontali, C. (1990). Long insertions within telomeres contribute to chromosome size polymorphism in *Plasmodium berghei*. *Mol. Cell. Biol.*, **10**, 6759–64.

Panyutin, I. G., Kovalsky, O. I., and Budowsky, E. I. (1989). Magnesium-dependent supercoiling-induced transition in $(dG)_n \cdot (dC)_n$ stretches and formation of a new G-structure by $(dG)_n$ strand. *Nucleic Acids Res.*, **17**, 8257–71.

Panyutin, I. G., Kovalsky, O. I., Budowsky, E. I., Dickerson, R. E., Rikhirev, M. E., and Lipanov, A. A. (1990). G-DNA: a twice-folded DNA structure adopted by single-stranded oligo(dG) and its implications for telomeres. *Proc. Natl. Acad. Sci. USA*, **87**, 867–70.

Perrot, M., Barreau, C., and Begueret, J. (1987). Nonintegrative transformation in the filamentous fungus *Podospora anserina*: stabilization of a linear vector by the chromosomal ends of *Tetrahymena thermophila*. *Mol. Cell. Biol.*, **7**, 1725–30.

Petracek, M. E. and Berman, J. (1992). *Chlamydomonas reinhardtii* telomere repeats form unstable structures involving guanine-guanine base pairs. *Nucleic Acids Res.*, **20**, 89–95.

Petracek, M. E., Lefebvre, P. A., Silflow, C. D., and Berman, J. (1990). *Chlamydomonas* telomere sequences are A+T-rich but contain three consecutive G·C base pairs. *Proc. Natl. Acad. Sci. USA*, **87**, 8222–26.

Pluta, A. F., Kaine, B. P., and Spear, B. B. (1982). The terminal organization of macronuclear DNA in *Oxytricha fallax*. *Nucleic Acids Res.*, **10**, 8145–54.

Ponzi, M., Pace, T., Dore, E., and Frontali, C. (1985). Identification of a telomeric DNA sequence in *Plasmodium berghei*. *EMBO J.*, **4**, 2991–6.

Ponzi, M., Janse, C. J., Dore, E., Scotti, R., Pace, T., Reterink, T. J. F., *et al.* (1990). Generation of chromosome size polymorphism during *in vivo* mitotic multiplication of *Plasmodium berghei* involves both loss and addition of subtelomeric repeat sequences. *Mol. Biochem. Parasitol.*, **41**, 73–82.

Ponzi, M., Pace, T., Dore, E., Picci, L., Pizzi, E., and Frontali, C. (1992). Extensive turnover of telomeric DNA at a *Plasmodium berghei* chromosomal extremity marked by a rare recombinational event. *Nucleic Acids Res.*, **20**, 4491–7.

Powell, W. A. and Kistler, H. C. (1990). *In vivo* rearrangement of foreign DNA by *Fusarium oxysporum* produces linear self-replicating plasmids. *J. Bacteriol.*, **172**, 3163–71.

Price, C. M. and Cech, T. R. (1987). Telomeric DNA-protein interactions of *Oxytricha* macronuclear DNA. *Genes Dev.*, **1**, 783–93.

Raghuraman, M. K. and Cech, T. R. (1990). The effect of monovalent cation-induced telomeric DNA structure on the binding of *Oxytricha* telomeric protein. *Nucleic Acids Res.*, **18**, 4543–52.

Renauld, H., Aparicio, O. M., Zierath, P. D., Billington, B. L., Chhablani, S. K., and Gottschling, D. E. (1993). Silent domains are assembled continuously from the telomere and are defined by promoter distance and strength, and by *SIR3* dosage. *Genes Dev.*, **7**, 1133–45.

Richards, E. J. and Ausubel, F. M. (1988). Isolation of a higher eukaryotic telomere from Arabidopsis thaliana. *Cell*, **53**, 127–36.

Richards, E. J., Goodman, H. M., and Ausubel, F. M. (1991). The centromere region of *Arabidopsis thaliana* chromosome 1 contains telomere-similar sequences. *Nucleic Acids Res.*, **19**, 3351–7.

Richards, E. J., Chao, S., Vongs, A., and Yang, J. (1992). Characterization of *Arabidopsis thaliana* telomeres isolated in yeast. *Nucleic Acids Res.*, **20**, 4039–46.

Röder, M. S., Lapitan, N. L. V., Sorrells, M. E., and Tanksley, S. D. (1993). Genetic and physical mapping of barley telomeres. *Mol. Gen. Genet.*, **238**, 294–303.

Sadhu, C., McEachern, M. J., Rustchenko-Bulgac, E.P., Schmid, J., Soll, D. R., and Hicks, J. B. (1991). Telomeric and dispersed repeat sequences in *Candida* yeast and their use in strain identification. *J. Bacteriol.*, **173**, 842–50.

Saiga, H. and Edström, J.E. (1985). Long tandem arrays of complex repeat units in *Chironomus* telomeres. *EMBO J.*, **4**, 799–804.

Scaria, P. V., Shire, S. J., and Shafer, R. H. (1992). Quadruplex structure of $d(G_3T_4G_3)$ stabilized by K^+ or Na^+ is an asymmetric hairpin dimer. *Proc. Natl. Acad. Sci. USA*, **89**, 10336–40.

Schechtman, M. G. (1990). Characterization of telomere DNA from *Neurospora crassa*. *Gene*, **88**, 159–65.

Schwarzacher, T. and Heslop-Harrison, J. S. (1991). *In situ* hybridization to plant telomeres using synthetic oligonucleotides. *Genome*, **34**, 317–23.

Sen, D. and Gilbert, W. (1988). Formation of parallel four-stranded complexes by guanine-rich motifs in DNA and its implications for meiosis. *Nature*, **334**, 364–6.

Sen, D. and Gilbert, W. (1990). A sodium-potassium switch in the formation of four-stranded G4-DNA. *Nature*, **344**, 410–14.

Sen, D. and Gilbert, W. (1992). Novel DNA superstructures formed by telomere-like oligomers. *Biochemistry*, **31**, 65–70.

Shampay, J. and Blackburn, E. H. (1988). Generation of telomere-length heterogeneity in *Saccharomyces cerevisiae*. *Proc. Natl. Acad. Sci. USA*, **85**, 534–8.

Shampay, J., Szostak, J. W., and Blackburn, E. H. (1984). DNA sequences of telomeres maintained in yeast. *Nature*, **310**, 154–7.

Simoens, C. R., Gielen, J., Van Montagu, M., and Inzé, D. (1988). Characterization of highly repetitive sequences of *Arabidopsis thaliana*. *Nucleic Acids Res.*, **14**, 6753–66.

Smith, F. W. and Feigon, J. (1992). Quadruplex structure of *Oxytricha* telomeric DNA oligonucleotides. *Nature*, **356**, 164–8.

Southern, E. M. (1970). Base sequence and evolution of guinea-pig α-satellite DNA. *Nature*, **227**, 794–8.

Stoll, S., Schmid, M., and Lipps, H. J. (1991). The organization of macronuclear DNA sequences associated with C_4A_4 repeats in the polytene chromosomes of *Stylonychia lemnae*. *Chromosoma*, **100**, 300–4.

Stoll, S., Zirlik, T., Maercker, C., and Lipps, H. J. (1993). The organization of internal telomeric repeats in the polytene chromosomes of the hypotrichous ciliate *Stylonychia lemnae*. *Nucleic Acids Res.*, **21**, 1783–8.

Sugawara, N. (1989). DNA sequences at the telomeres of the fission yeast *S. pombe*. Ph.D. Thesis, Harvard University.

Sugawara, N. and Szostak, J. W. (1986). Telomeres of Schizosaccharomyces pombe. *Yeast*, **2**, S373.

Sundquist, W. I. and Klug, A. (1989). Telomeric DNA dimerizes by formation of guanine tetrads between hairpin loops. *Nature*, **342**, 825–9.

Szostak, J. W. and Blackburn, E. H. (1982). Cloning yeast telomeres on linear plasmid vectors. *Cell*, **29**, 245–55.

Teschke, C., Solleder, G., and Moritz, K. B. (1991). The highly variable pentameric repeats of the AT-rich germline limited DNA in *Parascaris univalens* are the telomeric repeats of somatic chromosomes. *Nucleic Acids Res.*, **19**, 2677–84.

Van der Ploeg, L. H. T., Liu, A. Y. C., and Borst, P. (1984). Structure of the growing telomeres of Trypanosomes. *Cell*, **36**, 459–68.

Vernick, K. D. and McCutchan, T. F. (1988). Sequence and structure of a *Plasmodium falciparum* telomere. *Mol. Biochem. Parasitol.*, **28**, 85–94.

Vernick, K. D., Walliker, D., and McCutchan, T. F. (1988). Genetic hypervariability of telomere-related sequences is associated with meiosis in *Plasmodium falciparum*. *Nucleic Acids Res.*, **16**, 6973–85.

Walliker, D., Quakyi, I. A., Wellems, T. E., McCutchan, T. F., Szarfman, A., London, W. T., *et al.* (1987). Genetic analysis of the human malarial parasite *Plasmodium falciparum*. *Science*, **236**, 1661–6.

Walmsley, R. M., Chan, C. S. M., Tye, B.-K., and Petes, T. D. (1984). Unusual

DNA sequences associated with the ends of yeast chromosomes. *Nature*, **310**, 157–60.

Walsh, K. and Gualberto, A. (1992). MyoD binds to the guanine tetrad nucleic acid structure. *J. Biol. Chem.*, **267**, 13714–18.

Wang, S.-S. and Zakian, V. A. (1990). Sequencing of *Saccharomyces* telomeres cloned using T4 DNA polymerase reveals two domains. *Mol. Cell. Biol.*, **10**, 4415–19.

Wang, Y. and Patel, D. J. (1992). Guanine residues in d(T_2AG_3) and d(T_2G4) form parallel-stranded potassium cation stabilized G-quadruplexes with anti glycosidic torsion angles in solution. *Biochemistry*, **31**, 8112–19.

Wang, Y. and Patel, D. J. (1993). Solution structure of a parallel-stranded G-quadruplex DNA. *J. Mol. Biol.*, **234**, 1171–83.

Weiden, M., Osheim, Y. N., Beyer, A. L., and Van der Ploeg, L. H. T. (1991). Chromosome structure: DNA nucleotide sequence elements of a subset of the minichromosomes of the protozoan *Trypanosoma brucei*. *Mol. Cell. Biol.*, **11**, 3823–34.

Weisman-Shomer, P. and Fry, M. (1993). *QUAD*, a protein from hepatocyte chromatin that binds selectively to guanine-rich quadruplex DNA. *J. Biol. Chem.*, **268**, 3306–12.

Wellinger, R. J., Wolf, A. J., and Zakian, V. A. (1992). Use of non-denaturing Southern hybridization and two dimensional agarose gels to detect putative intermediates in telomere replication in *Saccharomyces cerevisiae*. *Chromosoma*, **102**, S150–S156.

Wellinger, R. J., Wolf, A. J., and Zakian, V. A. (1993*a*). Saccharomyces telomeres acquire single-strand TG_{1-3} tails late in S phase. *Cell*, **72**, 51–60.

Wellinger, R. J., Wolf, A. J., and Zakian, V. A. (1993*b*). Origin activation and formation of single-strand TG_{1-3} tails occur sequentially in late S phase on a yeast linear plasmid. *Mol. Cell. Biol.*, **13**, 4057–65.

Wells, R. A., Germino, G. G., Krishna, S., Buckle, V. J., and Reeders, S. T. (1990). Telomere-related sequences at interstitial sites in the human genome. *Genomics*, **8**, 699–704.

Williams, K., Doak, T. G., and Herrick, G. (1993). Developmental precise excision of *Oxytricha trifallax* telomere-bearing elements and formation of circles closed by a copy of the flanking target duplication. *EMBO J.*, **12**, 4593–601.

Williamson, J. R., Raghuraman, M. K., and Cech, T. R. (1989). Monovalent cation-induced structure of telomeric DNA: the G-quartet model. *Cell*, **59**, 871–80.

Wurster-Hill, D. H., Ward, O. G., Davis, B. H., Park, J. P., Moyzis, R. K., and Meyne, J. (1988). Fragile sites, telomeric DNA sequences, B chromosomes, and DNA content in raccoon dogs, *Nyctereutes procyonoides*, with comparative notes on foxes, coyote, wolf, and raccoon. *Cytogenet. Cell Genet.*, **49**, 278–81.

Yao, M.-C. and Yao, C.-H. (1981). Repeated hexanucleotide C-C-C-C-A-A is present near free ends of macronuclear DNA of *Tetrahymena*. *Proc. Natl. Acad. Sci. USA*, **78**, 7436–9.

Zahler, A. M., Williamson, J. R., Cech, T. R., and Prescott, D. M. (1991). Inhibition of telomerase by G-quartet DNA structures. *Nature*, **350**, 718–20.

Zakian, V. A. and Blanton, H. M. (1988). Distribution of telomere-associated sequences on natural chromosomes in *Saccharomyces cerevisiae*. *Mol. Cell. Biol.*, **8**, 2257–60.

4

Telomerase

In three ciliates and two mammals it has been shown that terminal repeat sequences are synthesized by a specialized DNA polymerase called telomerase (reviewed by Greider 1990; Blackburn 1992). This chapter investigates to what extent the known properties of telomerase are consistent with various features of the terminal repeat sequences of many species, including the majority for which telomerase has yet to be biochemically identified. These structural features have been described in detail in Chapter 3.

Identification of telomerase

Telomerase activity has been identified in five species. Although the biochemistry of the reaction has been analysed in depth, none of the protein components of any of these enzymes have been purified to homogeneity, and the genes encoding them have yet to be cloned.

Tetrahymena

Telomerase was first identified in this species and it remains one of the best characterized. The enzyme is present in vegetative cells but is found in higher amounts in cells undergoing macronuclear development, a time when there is a large *de novo* synthesis of telomeres (Greider and Blackburn 1985). *In vitro* it catalyses the addition of TTGGGG repeats one nucleotide at a time, rather than in blocks of the hexamer repeat, onto the 3' end of a range of single-stranded oligonucleotides. Oligonucleotides which prime telomerase correspond to the G-strand of *Tetrahymena, Saccharomyces, Euplotes*, human, and *Dictyostelium* terminal repeat sequences. Complementary C-strand oligonucleotides, such as $(CCCCAA)_4$, and a range of non-telomeric oligonucleotides are not utilized as primers (Greider and Blackburn 1985, 1987). Reaction products are usually radiolabelled, for example by including ^{32}P-labelled dGTP in the reaction mix, and are visualized on a sequencing gel as a ladder of products increasing in 1 nucleotide increments; there is also a characteristic 6 nucleotide periodicity in band intensity reflecting pausing at preferred sites in the repeat (see below). The reaction does not require any cofactors other than dGTP, dTTP, and the oligonucleotide.

Telomerase activity is sensitive to protease digestion, and the enzyme has an apparent molecular weight of 200–500 kDa as judged by column

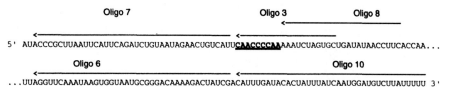

Fig. 4.1 *Tetrahymena* telomerase RNA template region defined by RNase H-directed cleavage. The 159 nt telomerase RNA component from *Tetrahymena* is shown. Oligonucleotide 3 in combination with RNase H abolishes telomerase activity *in vitro*; the other oligonucleotides have no effect. Oligonucleotide 8 also acts as a substrate for the enzyme. The putative template region is underlined. Arrowheads define the 3' end of each oligonucleotide. Reprinted with permission from *Nature* (Greider and Blackburn 1989), © Macmillan Magazines Ltd.

chromatography (Greider and Blackburn 1987). The activity is abolished by treatment with RNase A, and a 159 nucleotide RNA that co-purifies with *Tetrahymena* telomerase has been cloned and sequenced (Fig. 4.1). It contains the sequence 5'-CAACCCCAA-3', which is complementary to one and a half TTGGGG repeats (underlined in Fig. 4.1). Specific cleavage of various regions of this RNA has been accomplished by using RNase H in combination with a variety of DNA oligonucleotides complementary to short regions of the RNA. RNase H cleaves the RNA component of a DNA–RNA heteroduplex and this approach therefore directs cleavage to the region of the 159 nucleotide RNA complementary to the oligonucleotide. Telomerase with such a cleaved RNA can then be tested in a standard telomerase assay for activity. Only oligonucleotide 3 (see Fig. 4.1) in combination with RNase H abolishes telomerase activity. Because the only region of the RNA that appears to be sensitive to such cleavage contains the presumptive template sequence, this strongly implicates this region of the molecule in telomerase activity (Greider and Blackburn 1989; reviewed by Lamond 1989). Further evidence in support of the 5'-CAACCCCAA-3' motif providing the template is the observation that oligonucleotide 8 (Fig. 4.1), whose 3' end finishes at the nucleotide adjacent to the proposed template region, acts as a telomerase primer despite having none of the TG-rich features of other telomerase primers. Telomerase would therefore appear to be a specialized reverse transcriptase which catalyses DNA synthesis utilizing an integral RNA template.

Oxytricha

Telomerase activity has been found in crude macronuclear extracts of this ciliate. It adds TTTTGGGG repeats, one nucleotide at a time, onto the 3' end of oligonucleotides ending in repeats of this sequence. As with the other ciliate enzymes, it recognizes the 3' bases and adds the appropriate next base to continue the repeat (Zahler and Prescott 1988). This suggests an interaction between the template RNA and the 3' end of the primer, a point which is discussed in greater detail later. The reaction requires only the oligonucleotide, dGTP, and dTTP. It is sensitive to RNase treatment but the corresponding RNA has not been identified. The activity can also efficiently use as primers telomeric oligonucleotides complexed with the *Oxytricha* telomere binding protein (Shippen *et al.* 1994; Chapter 5).

Euplotes

Telomerase activity has also been identified in macronuclear extracts of this ciliate. It will add TTTTGGGG repeats, one nucleotide at a time, onto single-stranded oligonucleotides *in vitro* (Shippen-Lentz and Blackburn 1989). Oligonucleotides corresponding to the G-strand of *Euplotes, Saccharomyces, Dictyostelium,* and *Tetrahymena* function as primers for telomerase from this species. The 3' terminus of the primer is recognized and the next appropriate base is added to continue the repeat. The activity is sensitive to RNase treatment and the corresponding telomerase RNA has been identified and cloned. Although it is slightly longer, 192 nucleotides, and not very similar in sequence to the *Tetrahymena* RNA, it does contain a template for TTTTGGGG addition at a similar position in the sequence. Evidence that this region acts as a template for synthesis comes from the observation that only oligonucleotides which can pair with the RNA and whose 3' ends terminate in the presumptive template region act as telomerase primers (Shippen-Lentz and Blackburn 1990).

Humans

Although telomerase has been identified in three ciliate species, these organisms have a very unusual lifestyle involving chromosome breakage and *de novo* telomere addition to non-telomeric sequences, as discussed in detail in Chapter 6. It was therefore possible that telomerase is a specialized enzyme required for these unusual processes and is not involved in day-to-day telomere maintenance in other species. However, this does not appear to be the case, as telomerase activity has been identified in human and mouse cells.

Mammalian telomerase activity was first identified in a crude extract of human HeLa cells (Morin 1989). As might be expected from the relative number of telomeres, there is much less of this activity in a human cell than there is in *Tetrahymena*. It synthesizes TTAGGG repeats and can add over 60 repeats onto an oligonucleotide *in vitro*. The reaction is processive *in vitro*, as judged by an oligonucleotide challenge assay (Prowse *et al.* 1993). In addition to (TTAGGG)$_n$ a number of other oligonucleotides are utilized by human telomerase. These include oligonucleotides corresponding to *Oxytricha, Tetrahymena, Saccharomyces, Dictyostelium*, and *Arabidopsis* G-strands. It is not a weak, non-specific reaction with these heterologous repeat oligonucleotides, as many are used with efficiencies very similar to $(TTAGGG)_4$. A variety of non-telomeric or C-strand oligonucleotides do not function as primers. In an analogous fashion to the behaviour of ciliate telomerase, the 3′ ends of TTAGGG oligonucleotides are recognized and the next appropriate base is added to continue the repeat. A template RNA has not been identified but the telomerase activity is sensitive to RNase treatment, suggesting the presence of an essential RNA component. Telomerase activity has not been found in primary cells in culture but is detectable once they become immortal (Counter *et al.* 1992; see Chapter 7).

Mice

Telomerase activity has been identified in extracts of various mouse cell lines and is sensitive to RNase treatment, suggesting an essential RNA component (Prowse *et al.* 1993). It can utilize *Saccharomyces, Arabidopsis*, and *Tetrahymena* G-strand oligonucleotides as primers, as well as $(TTAGGG)_n$. Interestingly, under all conditions so far tested mouse telomerase appears to behave in a highly non-processive fashion, adding at most one or two TTAGGG repeats onto an oligonucleotide (Prowse *et al.* 1993) in contrast to the behaviour of human telomerase. It is curious that mice, which have much longer $(TTAGGG)_n$ arrays than humans (see Chapter 10), have a telomerase activity that synthesizes such short products *in vitro*. Telomere length therefore does not simply reflect the *in vitro* processivity of the enzyme. Processivity is discussed in detail in a following section, and the control of telomere length is discussed in Chapter 5.

The RNA component of telomerase has a conserved secondary structure

The telomerase template RNA has been cloned and sequenced from six *Tetrahymena* species, as well as the related species *Glaucoma* (Romero and Blackburn 1991). In each there is a motif complementary to the terminal repeat sequence which could function as a template. A 22 nucleotide region including the presumptive templating motif is highly conserved, but the remainder of the RNAs have no substantial regions of primary sequence similarity. However, a number of regions have the potential to base-pair and form stem–loop structures, and all the telomerase RNA molecules have the potential to form a similar secondary structure. Evidence that a specific secondary structure is adopted *in vivo* comes from covariance; for each species any nucleotide changes in positions that fall within a predicted stem–loop region are accompanied by compensatory changes elsewhere in the RNA to restore the base-pairing potential (Romero and Blackburn 1991; ten Dam *et al.* 1991). However, although the evolutionary studies suggest that the potential to form a secondary structure is conserved, there is no direct evidence as yet that such a structure is formed *in vivo* and little suggestion what its function might be. It would therefore be of interest to determine the effect of mutations which disrupt this secondary structure on telomerase function; this might be achieved by expressing variant RNA species *in vivo* (see below) or reconstituting telomerase with a mutant template RNA *in vitro* (Autexier and Greider 1994).

Primer selection by telomerase

In vitro telomerase shows a marked preference for certain single-stranded oligonucleotides as primers. How does telomerase select a DNA molecule onto which to add repeats; what features of a sequence are recognized and how is this recognition achieved? As described above, the RNA component of *Tetrahymena* telomerase contains a motif (5'-CAACCCCAA -3') that appears to provide the template for polymerization. However, there is no *a priori* reason for this sequence to have a role in the initial choice of primers for polymerization; this could occur by protein–DNA interaction, for example. Indeed, it has been shown that both *Tetrahymena* and human telomerases can elongate oligonucleotides whose 3' ends do not have extensive complementarity to the template region (Harrington and Greider 1991; Morin 1991).

The *Tetrahymena* enzyme will utilize oligonucleotides which terminate in TTGGGG at their 3' ends and non-telomeric oligonucleotides generally do not function as primers. However, a number of oligonucleotides which consist of pBR322 sequences at the 3' end will function as primers, but only if there is also a short stretch of TTGGGG repeats at the 5' end of the molecule. For example, one oligonucleotide which functions as a primer consists of two TTGGGG repeats at the 5' end followed by 36 bases of pBR322 sequence. This not only primes synthesis but the new repeats are added onto the end of the pBR322 sequences, arguing against the possibility that these oligonucleotides function because there is nucleolytic cleavage to remove the pBR322 sequences and reveal TTGGGG at the 3' terminus. Initiation on such chimeric oligonucleotides is of comparable efficiency to using a (TTGGGG)$_4$ primer. The final nucleotides of this oligonucleotide would be unable to pair with the template region, suggesting that pairing of the 3' terminus with the RNA is not essential for efficient use as a primer. Human telomerase appears to behave in a similar fashion. One case of human α-thalassaemia is caused by a truncation of chromosome 16, and (TTAGGG)$_n$ repeats have been added directly to the breakpoint despite little if any sequence similarity to TTAGGG at this point (Chapter 6). However, oligonucleotides corresponding to this region do function as primers for both human and *Tetrahymena* telomerase *in vitro* (Harrington and Greider 1991; Morin 1991) and recognition as a primer may reflect a TG-rich sequence 5' to the breakpoint. These lines of evidence indicate that primer selection can involve recognition of sequences many tens of nucleotides from the 3' terminus.

An *in vivo* example consistent with such a second site is reported by Barnett *et al.* (1993). During telomere-mediated chromosome fragmentation experiments in mammalian cells (see Chapter 10) it was observed that plasmid constructs lacking (TTAGGG)$_n$ were not detectably healed by (TTAGGG)$_n$ addition. However, constructs with a (TTAGGG)$_n$ tract terminating in about 30 bp of non-telomeric polylinker sequence were efficiently healed. Sequencing the healed products indicated that in some instances the *de novo* (TTAGGG)$_n$ addition had occurred directly onto polylinker sequence, rather than trimming back to the (TTAGGG)$_n$ tract and then sequence addition. Although it is not known if these healing events are mediated by telomerase, this *in vivo* example appears analogous to the utilization of pBR322 oligonucleotides with 5'-TTGGGG sequence by *Tetrahymena* telomerase described above. Another such example is the ability of yeast to add TG$_{1-3}$ repeats onto non-telomeric sequences provided that there are telomeric sequences within 30 nucleotides of the end of the molecule (Murray *et al.* 1988).

There is clearly some interaction between the 3' end of the primer and the template RNA. This can be shown by using oligonucleotides that terminate at different positions in the repeat. For example, oligonucleotides terminating in TTGGG, GGTT, or TTGG are respectively extended by *Tetrahymena* telomerase to produce TTGGG$\underline{GTTGGGG}$ ..., GGTT$\underline{GG}$-$\underline{GGTTGGGG}$ ..., and TTGG$\underline{GGTTGGGG}$ ..., (Greider and Blackburn 1987; Blackburn *et al.* 1989). Therefore the enzyme is able to recognize the 3' end of the oligonucleotide and add the next appropriate nucleotide.

The evidence therefore suggests that telomerase can interact with sequences both at the 3' terminus and more internally, which might occur in a number of ways. Firstly primer recognition could occur via pairing of a 5' sequence to the RNA template, and then the enzyme either slides along the primer to the 3' end or the intervening DNA loops out. Another possibility is that 5' sequences are recognized by an interaction with a protein component of telomerase. Detailed kinetic studies of *Tetrahymena* telomerase suggests sequence-specific interactions between 5' regions of the primer and the enzyme, separate from hybridization to the template RNA, which can have an effect on enzyme kinetics (Collins and Greider 1993; Lee and Blackburn 1993). This would be consistent with at least one component of the telomerase reaction involving protein–DNA interaction. The possibility of such an interaction may be of interest with respect to some known cases of *in vivo* telomere addition. In some species the process of healing of chromosome breaks is such that any sequence can be healed, irrespective of the presence of telomere-like sequence at or near the break (see Chapter 6). If the absence of telomere-like sequences at the 3′ end of the oligo-nucleotide can be overcome by a separate protein–DNA interaction, it is possible to speculate that a protein distinct from telomerase could promote an interaction between telomerase and chromosome ends which would not otherwise be recognized. Spangler *et al.* (1988) have cloned and sequenced a number of sites of telomere addition that arose during macronuclear development in *Tetrahymena*. The (TTGGGG)$_n$ repeats have been added onto A + T-rich DNA, but oligonucleotides corresponding to these telomere-adjacent sequences do not function as primers for *Tetrahymena* telomerase *in vitro*. Although there is no direct evidence, because *Tetrahymena* telo-merase appears to be able to interact with DNA ends *in vivo* which it does not recognize *in vitro*, this would be consistent with the speculation that there is an additional protein present in ciliates to facilitate the association of telomerase with broken ends consisting of non-telomeric sequences during macronuclear development. One might extend this speculation and suggest that in species without this specialized life cycle such a protein might not be present, thus preventing the efficient healing of random breaks.

Another possible criterion for use as a primer might in principle be the ability to form a fold-back G-quartet structure. However, as discussed in Chapter 3, the formation of such structures does not appear to be necessary for an oligonucleotide to function as a telomerase primer, and in fact if formed they actually appear to inhibit telomerase (Zahler *et al.* 1991). The observation that oligonucleotides that cannot form G-quartets, such as those with only a single TTGGGG or TTAGGG repeat, are able to function as primers further argues against G-quartet formation being required for primer function (Harrington and Greider 1991; Morin 1991).

Telomerase can synthesize telomeric sequences *in vivo*

There is strong evidence that telomerase is responsible for sequence addition to pre-existing *Tetrahymena* telomeres *in vivo*. This has been shown by transforming *Tetrahymena* with an altered version of the RNA component of telomerase and demonstrating that the predicted change in terminal repeat sequences occurs.

Cloned DNA can be introduced into *Tetrahymena* by microinjection, and Yu *et al.* (1990) have used this method to introduce the gene encoding the RNA component of *Tetrahymena* telomerase. By using a vector carrying the *cis*-acting replication control elements of the amplified rDNA mole-cules, the introduced gene is amplified and maintained at a much higher copy number than the endogenous gene, and a significant fraction of the cellular telomerase RNA is now produced from the introduced gene. The overall amount of telomerase RNA does not change, perhaps reflecting degradation of excess RNA not complexed with telomerase proteins.

Microinjection of a wild-type gene produced no obvious phenotype. However, three different mutant genes were also introduced and these did have an effect. The mutations led to RNA molecules being produced with changes in the template regions: these were an extra C (5'-CAACCCCCAA-3'), a C to T change (5'-CAACCTCAA-3'), and an A to G change (5'-CGACCCCAA-3'). All three mutant RNAs resulted in altered telomere structure, as judged by measuring telomere length on Southern blots using a (TTGGGG)₂ probe. The RNA with the 5'-CAACCTCAA-3' mutation was associated with a decrease in telomere length, whereas the other two correlated with an increase in telomere length (Yu *et al.* 1990).

Two of these altered RNAs also resulted in a change in telomere sequence. The predicted repeats were synthesized in the strains containing the 5'-CAACCCCCAA-3' and 5'-CGACCCCAA-3' mutant RNAs, as shown both by probing Southern blots with the corresponding oligonucleotide probes and by cloning and sequencing telomeric sequences from these strains. As a sequence change in the RNA template region results in the corresponding change in the telomere sequence *in vivo*, this is strong evidence that telomerase is involved in sequence addition to telomeres *in vivo* and that the RNA component is providing a template for repeat synthesis (Yu *et al.* 1990).

Interestingly, strains expressing the 5'-CAACCTCAA-3' RNA did not appear to synthesize any terminal repeats corresponding to this sequence, and introduction of this mutant RNA also resulted in shortened telomeres. It is possible that telomerase that has incorporated this mutant RNA is non-functional and that the cell is relying on the small amount of wild-type telomerase to maintain its telomeres. Despite not altering the sequence of its telomeres this strain showed gross morphological changes and eventual death of the cells in culture. One possible explanation is that the amount of functional telomerase is so low that there is progressive loss of telomeric sequences, eventually leading to loss of the entire terminal repeat array. However, it remains unclear how the change in telomere structure is linked to the eventual cell death. Neither is it clear why this mutant RNA does not produce a functional telomerase. It may be relevant that the base changed in this third mutant is subject to an uncharacterized modification in the wild-type RNA (Greider and Blackburn 1989), as shown by purified cellular telomerase RNA being resistant to cleavage by any RNase at three positions in the template region (underlined: 5'-CAACCCCAA-3'). It has been speculated that this modification has an essential role in the telomerase reaction and that the mutant RNA is not able to be modified, although there are few data regarding this (Blackburn 1992). Whatever the explanation, this result does indicate that there are sequence constraints on what terminal repeat sequence can be synthesized by telomerase; not all sequences in the template region lead to a functional enzyme. This may explain some of the conservation of terminal repeat sequences between species.

Strains containing the two mutant RNAs which did lead to new telomere sequences being synthesized were also sick, and the cultures eventually died. The exact reason for this is not clear but presumably cannot be the result of the loss of terminal sequence *per se*, because telomeric repeats are being maintained even though they are altered in sequence. Although this has not been demonstrated directly, it seems likely that over a period of time the wild-type terminal repeats become progressively replaced with mutant repeats. It is then possible to speculate that this may in turn cause the telomeres to lose their ability to bind a sequence-specific telomere binding protein, one unable to bind mutant repeats. Such a protein has not been identified biochemically in *Tetrahymena*, but it is valid to speculate that one might exist given the existence of highly sequence-specific telomere binding proteins in other ciliates (Chapter 5). This hypothetical *Tetrahymena* protein could perform an essential function such as protecting the chromosomes from end-to-end fusion or organizing the fragmented macronuclear genome into a higher-order structure as may occur in *Oxytricha*. One could then argue that the progressive replacement of wild-type terminal repeats by the mutant sequences correlates with a progressive loss of this protein bound at the telomeres and eventually cell death. Nevertheless, there is little direct evidence regarding the cause of the senescent phenotype in these cells.

Telomerase is responsible for programmed healing *in vivo*

The experiments described above involved the introduction of an altered telomerase RNA into vegetative cells, and demonstrated that telomerase was responsible for adding terminal repeats onto pre-existing telomeres. However, is telomerase responsible for the *de novo* addition of repeats onto non-telomeric sequences, such as that which occurs during programmed chromosome breakage? As mentioned above, during macronuclear development in *Tetrahymena* (TTGGGG)$_n$ repeats are added onto A + T-rich sequences with no similarity to the terminal repeat sequences, and which do not function as primers for telomerase *in vitro*. This, and a number of other examples where terminal repeats are added onto unrelated sequences (discussed in detail in Chapter 6), might suggest that telomerase is not responsible for telomere sequence addition in these cases. This possibility has been addressed by injecting mutant telomerase RNA genes into *Tetrahymena* cells which are then allowed to conjugate and undergo programmed genome rearrangement. The telomeres that are formed by *de novo* sequence addition onto non-telomeric sequences also carry the mutant terminal repeat sequences. The very first repeat added can be mutant, implicating telomerase in this *de novo* addition (Yu and Blackburn 1991).

Wild-type *Tetrahymena* usually produces very homogeneous arrays of (TTGGGG)$_n$. One interesting result from these experiments is that when telomeres are sequenced from cells expressing the 5'-CAACCCCCAA -3' mutant RNA, not only TTGGGGG repeats are found (as predicted from the template) but also TTGGGGGG, TTGGGGGGG, TTGGGGGGGG, and TTGGGGGGGGG repeats (Yu and Blackburn 1991). This seems to indicate that this version of telomerase inherently adds variable sequences, perhaps reflecting slippage of the product on the template RNA during synthesis. Such slippage during synthesis is one speculative explanation for the heterogeneous terminal repeats of *Saccharomyces* (TG_{1-3}) and *Dictyostelium* (AG_{1-8}), which vary from repeat to repeat in the number of Gs. This slippage does not reflect an inherent inability to synthesize accurately a repeat with more than four guanines, as *Cryptococcus neoformans* is capable of synthesizing homogeneous $(TTAGGGGG)_n$ repeats (Chapter 3).

Comparison of *in vivo* and *in vitro* processivity

Initial *in vitro* experiments with the *Tetrahymena* enzyme revealed that it synthesized repeats in a processive manner (Greider 1991). Once synthesis has started an enzyme molecule will generally continue to extend that product rather than dissociate. This can be shown by allowing synthesis to commence and then challenging with an excess of fresh oligonucleotide; if processive the enzyme will continue synthesis of partially completed products. However, subsequent studies have indicated that *Tetrahymena* telomerase can also behave in a non-processive fashion under certain conditions, such as when using oligonucleotides shorter than 10 nucleotides (Collins and Greider 1993; Lee and Blackburn 1993).

Is *Tetrahymena* telomerase processive *in vivo*? Telomeric sequences have been cloned from *Tetrahymena* cells expressing both wild-type and mutant telomerase RNAs (Yu *et al.* 1990; Yu and Blackburn 1991), and one such telomere had the following interspersion of wild-type (GGGGTT) and mutant (GGGGTC) repeat sequences:

$$5' \, (GGGGTT)_{30}(GGGGT\underline{C})_2GGGGTT(GGGGT\underline{C})_8GGGGTTGGGGTT\underline{C}(GGGGTT)_n \ldots 3'$$

Although this might be taken to imply a distributive telomerase action *in vivo*, it should be noted that these telomeres were obtained many generations after introduction of the mutant RNA gene, and thus many cycles of sequence addition and loss will have occurred, which could break up long stretches of one sequence, as could recombination between telomeres. Nevertheless, if a distributive mode does occur *in vivo* one might tentatively speculate that this reflects the G-strand of the telomere having a short single-stranded terminus *in vivo*, resulting in telomerase acting in the non-processive mode seen when using unusually short oligonucleotides *in vitro*.

In *Paramecium* the terminal repeat arrays are composed of a mixture of TTGGGG and TTTGGG repeats; *Plasmodium* arrays are composed of

TTTAGGG and TTCAGGG repeats (Chapter 3). The two repeats appear to be interspersed at random with each other, possibly the result of distributive synthesis *in vivo*. One speculative explanation for the presence of two variant repeats is that there are two types of telomerase in the cell, one for each repeat. If there are indeed two different telomerase activities present which were both processive, one would expect the production of arrays that contained long runs of one variant interspersed with long runs of the other. However, what is actually found is a high degree of interspersion, with very short runs of either variant. This may indicate that telomerase in these species is non-processive *in vivo*, a behaviour shown by the mouse enzyme *in vitro*. However, it is also possible that the absence of long stretches of homogeneity *in vivo* reflects instead the dynamic nature of telomeres; if there is constant loss and new sequence addition this could lead to breaking up of initially long and homogeneous runs of repeats and eventually lead to the repeat arrangement seen. Clarification of these points may require the isolation of telomerase from these species.

Although it is possible to perform detailed kinetic studies of telomerase action *in vitro*, this is a very artificial system. *In vitro* the substrate is a naked single-stranded oligonucleotide, whereas *in vivo* a macronuclear telomere may well be complexed with various proteins which affect the telomerase reaction. The environment in the macronucleus will also be very different. For example, there may be factors *in vivo* which do not co-purify with telomerase but affect its processivity; a precedent would be the protein PCNA, which is a processivity factor for DNA polymerase δ (Wang 1991). However, there is no precedent for a factor that causes an otherwise processive enzyme to act in a distributive fashion; all the identified accessory factors are required for processivity. Given the tentative nature of the data suggesting a distributive action *in vivo* and the artificial nature of the *in vitro* assay it is perhaps premature to compare *in vivo* and *in vitro* processivity.

A translocation step in the telomerase reaction

Tetrahymena telomerase is processive *in vitro*, with about 500 nucleotides being synthesized before half of the enzyme molecules dissociate. This processivity implies a translocation step in the telomerase reaction (Fig. 4.2). After synthesis of one repeat the 3′ end of the product must be repositioned with respect to the RNA template so that the next repeat can be synthesized. A distributive mode of action would involve the enzyme dissociating from the product, with the next repeat being synthesized by another enzyme molecule; in a processive reaction the same enzyme molecule synthesizes the next repeat. During repositioning any base-pairing between the product and the RNA template must be broken; however, some interaction between telomerase and the product must be maintained during repositioning so that this enzyme molecule can continue synthesis. One speculation is that such an interaction during repositioning is related to the interaction between telomerase and 5′ sequences in the primer suggested by, for example, the ability of *Tetrahymena* telomerase to utilize oligonucleotides terminating in pBR322 sequence as primers (see above). If processivity does indeed reflect two spatially separate interactions between the enzyme and the primer one might speculate that this provides an explanation for the non-processive elongation of short oligonucleotides; they might be too small for the 5′ region to have this interaction while the 3′ end is positioned at the site of polymerization (Collins and Greider 1993). One might also speculate that mouse telomerase acts in a non-processive manner because it lacks this second interaction. These possibilities must remain speculative in the absence of further evidence.

One unanswered question concerns the possibility that energy is required for repositioning. *Tetrahymena* telomerase requires only Mg^{2+}, dGTP, and dTTP for efficient extension of an oligonucleotide and has no requirement for ATP or GTP. As drawn in Fig. 4.2, repositioning requires the breaking of a number of base-pairs, which is likely to make it an energy-dependent step. However, telomerase RNA is not cleaved by RNase H during a reaction, suggesting that a stable DNA–RNA heteroduplex is not formed during synthesis and that the product may only be attached to the template RNA by a loose interaction of a few base-pairs (Greider and Blackburn 1989). Repositioning may therefore not require energy in the form of ATP hydrolysis.

Another prediction from the synthesis model shown in Fig. 4.2 is that the three most extreme 3′ nucleotides of the templating region (5′-CAACCC-CAA-3′) are not used as a template, functioning only in primer binding. *Tetrahymena* telomerase has been reconstituted using *in vitro* synthesized telomerase RNA, enabling various sequence changes in the RNA to be assayed for their effect on *in vitro* synthesis (Autexier and Greider 1994). Only mutations in the six most 5′ nucleotides of the template motif cause a change in the sequence synthesized by the enzyme, consistent with this model.

Synthesis by telomerase is not continuous *in vitro* but pauses at certain positions in the repeat. For example, the *Tetrahymena* enzyme appears to pause after addition of the first G in the TTGGGG repeat, resulting in every sixth band on the sequencing gel being more intense. The degree of pausing can be influenced by the relative dNTP concentrations, but some of it may reflect a pause which occurs while repositioning. It is suggested that the 5′-CAACCCCAA-3′ RNA template is copied to the 5′-most residue and then repositioning occurs (Greider 1991). If repositioning is slow relative to the rate of sequence addition then products will accumulate which terminate in TTG, as is observed.

When genomic DNA from ciliates (e.g. *Oxytricha*, *Stylonychia*, and

Telomerase

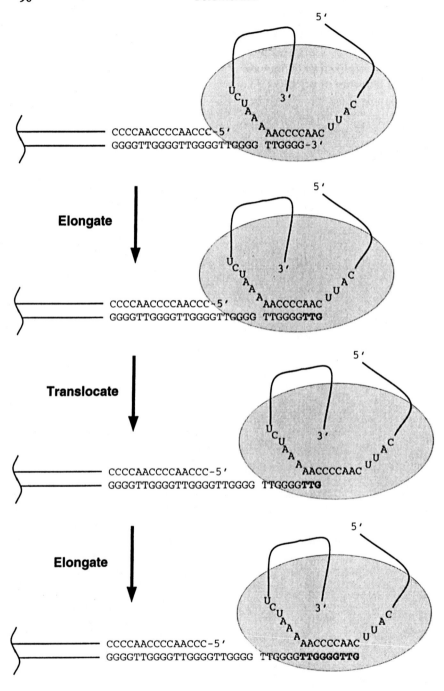

Euplotes) or slime moulds such as *Didymium* and *Physarum* is 3' end-labelled, Maxam–Gilbert chemical sequencing allows a clear sequence to be read (see Chapter 3). The success of such a reaction implies that most molecules terminate at the same position in the terminal repeat; for *Tetrahymena* this is . . . TTGGGG (Henderson and Blackburn 1989). One possibility is that this consistent terminal sequence reflects telomerase pausing predominantly at this position. However, *Tetrahymena* telomerase pauses after synthesis of . . . TTG *in vitro* and thus this does not explain the *in vivo* structure. Therefore some processing event after elongation by telomerase may be creating the defined end of *Tetrahymena* telomeres, such as a sequence-specific nuclease (Greider 1991). There is evidence for such a processing reaction in *Euplotes*, where during macronuclear development the terminal TTTTGGGG repeats are initially some 80 bp in length, shrinking to only 24 bp of duplex TTTTGGGG repeats in the mature macronucleus (Roth and Prescott 1985; Vermeesch and Price 1994). This shrinkage does not reflect losses because of the end-replication problem, as it also occurs in cells treated with aphidicolin to block DNA synthesis (Vermeesch *et al.* 1993). This suggests that the telomeres are initially synthesized in excess and are subsequently trimmed to their exact size and structure by a distinct processing reaction.

Telomerase also has a 3'–5' nucleolytic cleavage activity which occurs if the 3' end of a primer is paired with the C at the most 5' end of the template region, 5'-CAACCCCAA-3', the last position before repositioning (Collins and Greider 1993). The role of this cleavage reaction is far from clear, although it may reflect a mechanistic similarity of telomerase to the action of some RNA polymerases, which are able to cleave short transcripts and thus proceed through pause sites in the template (Kassavetis and Geiduschek 1993).

Does telomerase exist in *Saccharomyces cerevisiae*?

Telomerase has yet to be identified biochemically in *S. cerevisiae*. The similarity between yeast terminal repeats and those of species in which it has been shown that telomerase exists is one argument in favour of a yeast telomerase. The phenotype of the *EST1* mutation is consistent with the *EST1* gene encoding a protein component of telomerase (see Chapter 5). However, recombination can occur between yeast telomeres during the healing of introduced heterologous terminal repeats by addition of yeast

---

Fig. 4.2 Model for sequence addition by telomerase. Telomerase recognizes the substrate, elongates the 3' terminus by copying the RNA template, then becomes repositioned with respect to the substrate to initiate a further round of synthesis. Reprinted with permission from *Nature* (Greider and Blackburn 1989), © Macmillan Magazines Ltd.

TG_{1-3} repeats, although it is far from clear whether this is a specialized reaction that occurs only during healing or whether it reflects a recombination reaction that is responsible for maintaining the terminal repeat sequences (see Chapter 6).

In the later part of S phase, immediately after the telomeric sequences have been replicated, at least some terminal TG_{1-3} arrays become single-stranded (Wellinger *et al.* 1992, 1993*a,b*). This has been shown by a modified Southern blot protocol which involves no denaturation of the genomic DNA. The blots are probed with strand-specific probes using a protocol capable of detecting single-stranded regions as small as 30 nucleotides. This demonstrated that some TG_{1-3} but none of the $C_{1-3}A$ strands of the terminal arrays become single-stranded for a short period in late S phase, immediately after replication. Such behaviour is not shown by chromosome-internal TG_{1-3} arrays. This would be consistent with elongation of the TG_{1-3} strand by a telomerase activity in yeast. However, as it has not been shown that there is any sequence addition to the G-rich strand (there is no evidence that this strand increases in length when it becomes single-stranded) it is difficult to rule out the possibility that this phenomenon does not simply reflect the nature of lagging strand DNA synthesis. As Okazaki fragments are around 200 nucleotides in length, this implies that primase synthesizes RNA primers spaced roughly this distance apart. For example, if an RNA primer is not synthesised within a 50 nucleotide internal region it will still be replicated almost immediately by extension of an Okazaki fragment from outside the region. However, for a 50 nucleotide region at the very end of the chromosome, lagging strand synthesis is dependent on extension of an RNA primer produced within it. If, when the replication fork passes through the telomere, it does not synthesize a primer within the last 50 nucleotides of the chromosome, there may well be a delay until one is synthesised. During this delay the other strand will remain single-stranded. For a yeast chromosomal telomere the implication is that the TG_{1-3} strand will be completely synthesized by leading strand synthesis, but the spacing of RNA primers may mean that there is a delay in synthesizing the complementary $C_{1-3}A$ during S phase. The consequence of this would be that the TG_{1-3} strand would appear to become single-stranded for a short time during S phase.

A number of sequences can be healed by the addition of yeast TG_{1-3} repeats *in vivo*. These include TTTTGGGG, TTGGGG, TTAGGG, and TTTAGGG repeats (Chapter 3). The ability of such repeats to promote the addition of TG_{1-3} repeats is similar to the ability of telomerase to utilize heterologous telomeric sequences as primers *in vitro*, and is consistent with a yeast telomerase healing these sequences *in vivo*. Perhaps the strongest evidence for a yeast telomerase is the observation that *de novo* telomere formation initiates with a very limited subset of repeats (Kramer and Haber 1993). They analysed 44 examples of telomere formation at short

TG-rich sequences in yeast, and found that in 37 cases one of two short motifs (GTGTGGGTGTG or GTGTGTGGGTGTG) were either the first nucleotides added or overlapped the junction between the chromosomal primer and the newly added sequences, consistent with a model where yeast telomerase activity adds GTGTGGGTGTG or GTGTGTGGGTGTG repeats.

Summary

At the end of Chapter 3 a number of structural features common to many telomeric sequences were listed.

1. Most terminal repeats are of the form $T_x G_y$ or $T_x AG_y$, but an important point illustrated in Table 3.1 is that terminal repeat sequences do vary between species. However, there seem to be sequence constraints, with most changes being variations in the *number* of thymines or guanines, rather than changing the bases themselves. Such constraints are suggested by the example of species as diverse as fungi, slime moulds, and mammals having TTAGGG as their repeat. A simplistic model is that any sequence could be synthesized by telomerase given the correct RNA template. However, there may be biochemical constraints on what sequences can be efficiently synthesized by telomerase. For example, a single base change in the template region of the *Tetrahymena* RNA apparently abolishes telomerase action. Another possibility is that any sequence could in principle be synthesized by telomerase, but only certain sequences can fulfil the other requirements for a telomeric sequence, such as protein binding or possibly formation of some G-quartet or other structure.

2. As a direct consequence of telomerase synthesis of the DNA strand with its 3′ terminus at the end of the chromosome the G-strand has a fixed orientation. The asymmetry in base composition reflects the sequence of the repeat synthesized by telomerase.

3. The primer gap (Fig. 1.1) caused by incomplete replication is sufficient to make the terminus of the G-rich strand single-stranded, although telomerase would also have this effect (Fig. 1.5).

4. Heterogeneity in the size of the terminal repeat array could be caused by variation in the number of repeats added by telomerase. However, loss due to incomplete replication is in itself sufficient to produce a smearing of terminal restriction fragments, as demonstrated by broken chromosome ends in *Drosophila* (Chapter 9). The known properties of telomerase are insufficient to account for the exact length and structure

of ciliate macronuclear telomeres, and this may indicate a separate processing reaction to produce these ends.

5. The greater rate of sequence turnover in the terminal region of the repeat arrays is consistent with incomplete replication and telomerase addition being processes which act at the end of the molecule. However, gene conversion events would also be predicted to add sequence to the end of the molecule (Fig. 1.3) and thus it is not possible to discriminate between the two mechanisms on this basis.

6. The heterogeneous terminal repeats of species such as *Saccharomyces* (TG_{1-3}) and *Dictyostelium* (AG_{1-8}) have been speculated to reflect an intrinsic property of telomerase, although there is no direct evidence to support this. This follows from the observation that an altered version of *Tetrahymena* telomerase is capable of slippage during repeat synthesis, resulting in variation in the sequence of the repeats synthesized.

7. The presence of two repeat sequences interspersed with each other in species such as *Plasmodium* and *Paramecium* has been speculated to reflect two distinct telomerases, each with a different template sequence, although again there is no direct evidence for this.

Although as yet only a few species have been shown to possess a telomerase activity this may be largely because few other species have been investigated biochemically. A telomerase model can rationalize, or at least be consistent with, many of the observed structural features of telomeres. Most terminal repeats conform to the form $T_x G_y$ or $T_x AG_y$, and it would be reasonable to suggest that the existence of telomerase in one species may imply the existence of a similar activity in other species with the same or similar repeat sequence.

References

Reviews

Blackburn, E. H. (1992). Telomerases. *Annu. Rev. Biochem.*, **61**, 113–29.

Greider, C. W. (1990). Telomeres, telomerase and senescence. *Bioessays*, **12**, 363–9.

Kassavetis, G. A. and Geiduschek, E. P. (1993). RNA polymerase marching backward. *Science*, **259**, 944–5.

Lamond, A. I. (1989). *Tetrahymena* telomerase contains an internal RNA template. *Trends Biochem. Sci.*, **14**, 202–4.

Wang, T. S.-F. (1991). Eukaryotic DNA polymerases. *Annu. Rev. Biochem.*, **60**, 513–52.

Primary papers

Autexier, C. and Greider, C. W. (1994). Functional reconstitution of wild-type and mutant *Tetrahymena* telomerase. *Genes Dev.* **8**, 563–75.

Barnett, M. A., Buckle, V. J., Evans, E. P., Porter, A. C. G., Rout D., Smith, A. G., and Brown, W. R. A. (1993). Telomere directed fragmentation of mammalian chromosomes. *Nucleic Acids Res.*, **21**, 27–36.

Blackburn, E. H., Greider, C. W., Henderson, E., Lee, M. S., Shampay, J., and Shippen-Lentz, D. (1989). Recognition and elongation of telomeres by telomerase. *Genome*, **31**, 553–60.

Collins, K. and Greider, C. W. (1993). *Tetrahymena* telomerase catalyzes nucleolytic cleavage and nonprocessive elongation. *Genes Dev.*, **7**, 1364–76.

Counter, C. M., Avilion, A. A., LeFeuvre, C. E., Stewart, N. G., Greider, C. W., Harley, C. B., and Bacchetti, S. (1992). Telomere shortening associated with chromosome instability is arrested in immortal cells which express telomerase activity. *EMBO J.*, **11**, 1921–9.

Greider, C. W. (1991). Telomerase is processive. *Mol. Cell. Biol.*, **11**, 4572–80.

Greider, C. W. and Blackburn, E. H. (1985). Identification of a specific telomere terminal transferase activity in Tetrahymena extracts. *Cell*, **43**, 405–13.

Greider, C. W. and Blackburn, E. H. (1987). The telomere terminal transferase of Tetrahymena is a ribonucleoprotein enzyme with two kinds of primer specificity. *Cell*, **51**, 887–98.

Greider, C. W. and Blackburn, E. H. (1989). A telomeric sequence in the RNA of *Tetrahymena* telomerase required for telomere repeat synthesis. *Nature*, **337**, 331–7.

Harrington, L. A. and Greider, C. W. (1991). Telomerase primer specificity and chromosome healing. *Nature*, **353**, 451–4.

Henderson, E. R. and Blackburn, E. H. (1989). An overhanging 3′ terminus is a conserved feature of telomeres. *Mol. Cell. Biol.*, **9**, 345–8.

Kramer, K. M. and Haber, J. E. (1993). New telomeres in yeast are initiated with a highly selected subset of TG_{1-3} repeats. *Genes Dev.*, **7**, 2345–56.

Lee, M. S. and Blackburn, E. H. (1993). Sequence-specific DNA primer effects on telomerase polymerization activity. *Mol. Cell. Biol.*, **13**, 6586–99.

Morin, G. B. (1989). The human telomere terminal transferase enzyme is a ribonucleoprotein that synthesizes TTAGGG repeats. *Cell*, **59**, 521–9.

Morin, G. B. (1991). Recognition of a chromosome truncation site associated with α-thalassaemia by human telomerase. *Nature*, **353**, 454–6.

Murray, A. W., Claus, T. E., and Szostak, J. W. (1988). Characterization of two telomeric DNA processing reactions in *Saccharomyces cerevisiae*. *Mol. Cell. Biol.*, **8**, 4642–50.

Prowse, K. R., Avilion, A. A., and Greider, C. W. (1993). Identification of a nonprocessive telomerase activity from mouse cells. *Proc. Natl. Acad. Sci. USA*, **90**, 1493–7.

Romero, D. P. and Blackburn, E. H. (1991). A conserved secondary structure for telomerase RNA. *Cell*, **67**, 343–53.

Roth, M. and Prescott, D. M. (1985). DNA intermediates and telomere addition during genome reorganization in Euplotes crassus. *Cell*, **41**, 411–7.

Shippen, D. E., Blackburn, E. H., and Price, C. M. (1994). DNA bound by the *Oxytricha* telomere protein is accessible to telomerase and other DNA polymerases. *Proc. Natl. Acad. Sci. USA*, **91**, 405–9.

Shippen-Lentz, D. and Blackburn, E. H. (1989). Telomere terminal transferase activity from *Euplotes crassus* adds large numbers of TTTTGGGG repeats onto telomeric primers. *Mol. Cell. Biol.*, **9**, 2761-4.

Shippen-Lentz, D. and Blackburn, E. H. (1990). Functional evidence for an RNA template in telomerase. *Science*, **247**, 546-52.

Spangler, E. A., Ryan, T., and Blackburn, E. H. (1988). Developmentally regulated telomere addition in *Tetrahymena thermophila*. *Nucleic Acids Res.*, **16**, 5569-85.

ten Dam, E., van Belkum, A., and Pleij, K. (1991). A conserved pseudoknot in telomerase RNA. *Nucleic Acids Res.*, **19**, 6951.

Vermeesch, J. R. and Price, C. M. (1994). Telomeric DNA sequence and structure following *de novo* telomere synthesis in *Euplotes crassus*. *Mol. Cell. Biol.*, **14**, 554-66.

Vermeesch, J. R., Williams, D., and Price, C. M. (1993). Telomere processing in *Euplotes*. *Nucleic Acids Res.*, **21**, 5366-71.

Wellinger, R. J., Wolf, A. J., and Zakian, V. A. (1992). Use of non-denaturing Southern hybridization and two dimensional agarose gels to detect putative intermediates in telomere replication in *Saccharomyces cerevisiae*. *Chromosoma*, **102**, S150-56.

Wellinger, R. J., Wolf, A. J., and Zakian, V. A. (1993a). Saccharomyces telomeres acquire single-strand TG_{1-3} tails late in S phase. *Cell*, **72**, 51-60.

Wellinger, R. J., Wolf, A. J., and Zakian, V. A. (1993b). Origin activation and formation of single-strand TG_{1-3} tails occur sequentially in late S phase on a yeast linear plasmid. *Mol. Cell. Biol.*, **13**, 4057-65.

Yu, G.-L. and Blackburn, E. H. (1991). Developmentally programmed healing of chromosomes by telomerase in Tetrahymena. *Cell*, **67**, 823-32.

Yu, G.-L., Bradley, J. D., Attardi, L. D., and Blackburn, E. H. (1990). *In vivo* alteration of telomere sequences and senescence caused by mutated *Tetrahymena* telomerase RNAs. *Nature*, **344**, 126-32.

Zahler, A. M. and Prescott, D. M. (1988). Telomere terminal transferase activity in the hypotrichous ciliate *Oxytricha nova* and a model for replication of the ends of linear DNA molecules. *Nucleic Acids Res.*, **16**, 6953-72.

Zahler, A. M., Williamson, J. R., Cech, T. R., and Prescott, D. M. (1991). Inhibition of telomerase by G-quartet DNA structures. *Nature*, **350**, 718-20.

5

Telomere proteins

Telomere function requires not only telomeric DNA sequences but also various interacting proteins such as telomerase (reviewed by Zakian 1989; Blackburn 1991; Henderson and Larson 1991). Protecting a telomere from fusion and recombination may not be a direct result of the DNA sequence but rather a result of the action of telomere binding proteins. In *S. cerevisiae* the RAP1 protein, in concert with a range of other factors, is speculated to form a heterochromatin-like domain at the telomere, making the double-stranded terminus inaccessible to nucleases or recombination enzymes as well as ensuring that natural chromosome ends do not activate cell cycle checkpoints which monitor DNA damage (see Chapter 2). Other telomere functions which may well reflect the activity of telomere binding proteins include physical interactions with the nuclear envelope and telomere–telomere interactions. Although almost nothing is known of the molecular nature of the interactions with the nuclear envelope (discussed in Chapter 2) the role of telomere binding proteins in the interactions between ciliate telomeres has been studied in some detail.

Some telomere proteins have been identified by their ability to bind telomeric DNA sequences directly *in vitro*. Such telomere binding proteins have been biochemically isolated from very few species, in contrast with the very wide range of species for which telomeric DNA sequences have been obtained (see Chapter 3). Some of the first evidence for the existence of telomere binding proteins came from unusual telomeric chromatin structure as judged by nuclease digestion. Indeed, such data remain the only direct evidence for telomere binding proteins in some species, such as *Dictyostelium* (Edwards and Firtel 1984) and *Tetrahymena* (Blackburn and Chiou 1981; Budarf and Blackburn 1986). However, as telomere structure is to a large extent conserved it is hoped that information regarding the biochemistry of telomere binding proteins from one species might have relevance to others. Some of the most extensively biochemically characterized telomere binding proteins are those of ciliates. These appear to bind the end of the DNA molecule, rather than coating the length of the terminal repeats, and their identification has been facilitated by the enormous number of telomeres in a ciliate macronucleus, about 4×10^7 for hypotrichous ciliates such as *Oxytricha nova*. Together with, for example, the ability to purify the *Euplotes* protein in essentially 1:1 stoichiometric amounts with chromosome ends, this has led to the purification of a large amount of protein from these species. In contrast, the task of identifying analogous proteins which bind to the terminus of mammalian chromosomes has proven to be much more difficult; for example, a mouse cell contains only 84 telomeres. Genetic methods have also been used in *S. cerevisiae* to isolate proteins which have a role in telomere biology. The advantage of a genetic approach is that it is also able to identify proteins involved in telomere function but which do not bind telomeric DNA sequences directly and the phenotypes of mutants can be investigated.

Biochemically identified proteins

Telomere binding proteins fall primarily into two categories: those that bind the single-stranded G-strand overhang and those that bind duplex DNA. The affinities of the various reported telomere binding proteins can be compared with those known sequence-specific DNA binding proteins. For example, a sequence-specific transcription factor would generally be expected to have a dissociation constant (K_D) in the range of 10^{10}–10^{12} M^{-1} for its interaction with its specific target sequence, falling to 10^6 for non-specific DNA binding, or for a non-specific DNA binding protein such as *E. coli* single strand binding protein (SSB). The cytoplasmic intermediate filament vimentin will bind oligonucleotides containing a number of telomere repeats including *Oxytricha*, yeast, human, and *Tetrahymena* G-strands (Shoeman *et al*. 1988; Shoeman and Traub 1990). However, only one oligonucleotide was found to which vimentin would not bind, and the K_D for the interaction of mouse vimentin with TTAGGG repeats is about $3 \times 10^7 M^{-1}$. These two observations may suggest that the observed DNA binding is non-specific, perhaps with some preference for base composition, and *a priori* it would seem unlikely that this cytoplasmic protein has any role in telomere function.

Oxytricha nova

This ciliate shares the same TTTTGGGG terminal repeat sequence as *Euplotes crassus*. The *Oxytricha* telomere binding protein is a heterodimer of a 41 kDa α subunit and a 56 kDa β subunit. Genes for both subunits have been cloned using partial sequence data from the purified proteins (Hicke *et al*. 1990; Gray *et al*. 1991). The heterodimer binds in a salt-stable fashion to the single-stranded G-strand terminus, protecting it from exonucleases and chemical modification (Gottschling and Zakian 1986). It cannot bind oligonucleotides that are fully folded into a fold-back G-quartet structure (Raghuraman and Cech 1990) and any that form may be removed by the β subunit (see Chapter 3).

The two subunits cannot be separated without destroying DNA binding activity and study of the individual binding properties required the cloning of their genes and expression of the two subunits in *E. coli* (Price and Cech

1989; Gray *et al*. 1991). The α subunit can bind DNA alone, and is able to bind internal TTTTGGGG tracts flanked by non-telomeric sequences. However, production of a methylation footprint similar to that seen *in vivo* requires the presence of the β subunit in a 1:1 ratio (Raghuraman and Cech 1989; Gray *et al*. 1991). In the presence of DNA the α and β subunits produce a ternary complex in a 1:1:1 ratio, but in its absence are found as monomers (Fang and Cech 1993a; Fang *et al*. 1993). The β subunit has two separable domains, one involved in promoting G-quartet formation (Chapter 3) and the other in dimerization with the α subunit (Hicke *et al*. 1990).

Under certain salt concentrations the protein–DNA complex forms high molecular weight aggregates. DNA bound in these aggregates has a methylation footprint more similar to that seen *in vivo* than that of the monomeric complex (Price and Cech 1987; Raghuraman and Cech 1989; Raghuraman *et al*. 1989; Gray *et al*. 1991). This suggests that a similar aggregation may be occurring *in vivo*, and that the *Oxytricha* protein may be mediating *in vivo* telomere–telomere associations.

Purified *Oxytricha* macronuclear DNA will cohere *in vitro* via G-tetrad structures to form aggregates (see Chapter 3) and similarly if *Oxytricha* are lysed in high salt the DNA is seen as aggregates on agarose gels (Lipps *et al*. 1982). The crucial difference is that the aggregates from the lysed cells are sensitive to protease digestion. Similarly, if *Holosticha* is lysed and its macronuclear DNA spread and viewed by electron microscopy, clusters of macronuclear chromosomes can be seen, held together in a rosette by their telomeres (Prescott 1983). However, these clusters are again abolished by protease digestion. Furthermore, the DMS footprint of *Oxytricha* telomeres *in vivo* (Price and Cech 1987) differs from that seen if the terminal repeats are folded into G-tetrad structures *in vitro* (Fang and Cech 1993b). Therefore, although purified DNA may cohere by the formation of G-tetrad structures *in vitro*, there is no evidence that G-tetrad structures form *in vivo*. These observations suggest instead that *in vivo* aggregation reflects protein–protein interactions.

What might be the *in vivo* role for such an aggregation? Ciliates face a major problem by fragmenting their macronuclear genome into small, kilobase-sized pieces. The genome of most eukaryotes is attached in loops to a fixed nuclear scaffold or matrix, the site of both replication and transcription (Jackson 1991). The ability to introduce superhelical stress into DNA may have an important role in the regulation of DNA replication and gene expression. If such a topologically constrained genome is important, then ciliates face a potential problem if the small macronuclear chromosomes are free to rotate. It has been speculated that they prevent this by converting their fragmented macronuclear genome into large topologically constrained molecules by telomere–telomere interactions (Fang and Cech 1993b). Such interactions might need to be very strong to withstand superhelical stress, which could explain why the ciliate telomere binding proteins bind DNA in such an extremely stable fashion. The hypothesis that *in vivo* interactions do occur is supported by the observation that chromatin in the macronucleus, when visualized by electron microscopy, is not found as short 30 nm chromatin fragments equivalent in size to the small molecules in purified DNA. Instead much longer 30 nm chromatin strands are seen, and their size is such that they must be formed by the end-to-end association of multiple gene-sized molecules (Meyer and Lipps 1981). The observation that the aggregation of the small macronuclear chromosomes *in vivo* appears to be protease-sensitive (Lipps *et al.* 1982; Prescott 1983) suggests that organization into long chromatin fibres is mediated primarily by protein–protein interactions rather than reflecting end-to-end DNA fusions. These *in vivo* interactions may be related to the aggregation of telomere DNA–protein complexes *in vitro*.

Euplotes crassus

The *Oxytricha* and *Euplotes* proteins behave similarly. Both protect telomeric DNA from nuclease digestion or chemical modification. They bind very tenaciously, yet non-covalently, in a highly salt-stable manner, with the proteins remaining complexed even in 2 M NaCl or 6 M CsCl. Indeed, the *Euplotes* and *Oxytricha* telomere binding proteins are essentially the only proteins that remain attached to chromosomal DNA under such conditions, which allows for the rapid isolation of large amounts of pure protein (Price 1990). The protein can only be removed from the DNA by denaturing the protein or removing the DNA with micrococcal nuclease. The extreme salt-stability of this protein–DNA interaction is unusual and may reflect a markedly different mode of DNA binding than of other proteins studied to date, and for this reason it would be of interest to know more about the nature of the binding site. However, salt-stable interactions are not unique to ciliate telomere binding proteins; *E. coli* SSB, which remains attached to poly(dT) even in 5 M NaCl, is another very salt-stable single-stranded DNA-binding protein (Lohman and Overman 1985). As discussed above, the extremely strong binding reaction observed with the ciliate telomere binding proteins may reflect a need to constrain the short macronuclear chromosomes topologically.

The purified *Euplotes* protein, a 51 kDa monomer, binds the G-strand of the *Euplotes* telomere when single-stranded, but it does not bind duplex DNA nor the C-strand. The natural *Euplotes* ends are of the form $(GGGGTTTT)_n GG$, and an oligonucleotide of this form binds the protein well. However, a similar oligonucleotide lacking the two terminal Gs does not bind the protein (Price *et al.* 1992). That it binds to the very end of the DNA as a 'cap' is supported by the purification of the protein in roughly 1:1 stoichiometric yields with chromosome ends (Price *et al.* 1992).

Two genes have been identified by screening a *Euplotes* genomic library with the cloned *Oxytricha* α subunit gene as a probe under reduced hybridization stringency (Wang *et al*. 1992). One gene encodes a 51 kDa protein which appears to be the telomere binding protein, based on identity to partial sequence data from the purified protein. The other gene encodes a 53 kDa polypeptide that is about 55% identical to the 51 kDa protein and has an as yet unknown biological role. It does not seem to be a pseudogene as it is transcribed and has a long open reading frame with the usual translational start and stop signals. As the 53 kDa protein is expressed during DNA replication and is localized to the replication bands, it may function during telomere replication (C. Price, personal communication).

The predicted amino acid sequence of both the *Euplotes* proteins are about 55% identical to the α subunit proteins from *Oxytricha* and *Stylonychia*. The main region of similarity is the N-terminal two-thirds of the protein. The N-terminal region of the 51 kDa *Euplotes* telomere binding protein probably contains most of the DNA binding activity since a 35 kDa N-terminal fragment of the *Euplotes* protein has the same sequence-specific and salt-stable binding as the entire protein (Price *et al*. 1992). The ability of the full-length protein to bind only terminal repeats is one feature where the *Euplotes* protein differs from the *Oxytricha* protein, although the 35 kDa N-terminal *Euplotes* peptide fragment has less of a preference for the extreme 3′ terminus and will also bind oligonucleotides with internal tracts.

The main difference between the *Euplotes* and *Oxytricha* proteins is that the latter is a heterodimer. However, although an equivalent of the *Oxytricha* β subunit has not been identified biochemically, despite the tight association of the two subunits in *Oxytricha*, there are DNA fragments in the *Euplotes* genome which cross-hybridize with the cloned *Oxytricha* β subunit gene. Until the corresponding loci have been cloned the presence of a β subunit homologue in *Euplotes* remains speculative (Wang *et al*. 1992).

Stylonychia

This ciliate is closely related to *Oxytricha* and has the same TTTTGGGG terminal repeat sequence. Genes encoding proteins 79% and 77% identical to the α and β subunits of the *Oxytricha* telomere binding protein have been identified by cross-hybridization to the cloned *Oxytricha* genes (Fang and Cech 1991). These proteins have not yet been biochemically characterized.

Saccharomyces cerevisiae

RAP1

This protein is known by a variety of names (RAP1, TUF, TBA, and GRF1) because it was independently identified in a number of laboratories. It is a DNA binding protein which plays a role in repression or activation of transcription depending on the context of its binding site. It binds to the transcriptional silencer elements that regulate expression of the silent mating-type loci (Shore and Nasmyth 1987; Buchman *et al.* 1988*a,b*; see Chapter 8). It also binds to the activation elements of some glycolytic and ribosomal protein genes (Huet and Sentenac 1987; Shore and Nasmyth 1987; Vignais *et al.* 1987; Buchman *et al.* 1988*a,b*; Chambers *et al.* 1989; Moehle and Hinnebusch 1991). It is involved in the unusually efficient segregation of circular plasmids bearing telomere repeat sequence tracts (Longtine *et al.* 1993; see Chapter 8). Finally, binding of RAP1 to a site upstream of the *HIS4* gene is necessary for the high rate of meiotic (but not mitotic) recombination seen at this locus; mutation of the RAP1 binding site reduces gene conversion, whereas overexpression of the RAP1 protein increases recombination to above wild-type levels (White *et al.* 1991).

The GT-rich consensus RAP1 binding site prompted an investigation of its role, if any, at telomeres. RAP1 does indeed bind the irregular TG_{1-3} terminal repeat tracts, at sites found on average every 18 bp (Buchman *et al.* 1988*b*; Gilson *et al.* 1993*b*; Graham and Chambers 1994). Furthermore, Berman *et al.* (1986) isolated a telomere binding protein using affinity columns containing TG_{1-3} DNA which was later identified as RAP1 (Longtine *et al.* 1989).

The gene encoding RAP1 has been cloned and is essential for viability (Shore and Nasmyth 1987). The cause of the lethality is not known, but may be because of its effect on the transcription of one or more genes encoding essential metabolic enzymes or ribosomal proteins. The *RAP1* gene is distinct from those identified by *tel1*, *tel2*, *est1*, or *cdc17* mutations. Temperature-sensitive alleles of the *RAP1* gene exist; strains with such *rap1[ts]* mutations grow well at 23°C but arrest at 37°C. At a semi-permissive temperature (around 30°C) they show a much reduced growth rate but do divide. It is thought that at this temperature levels of active RAP1 protein are reduced, but not sufficiently to be lethal. The effect of growing *rap1[ts]* strains at a semi-permissive temperature is that over a period of about 100 cell generations the terminal TG_{1-3} repeats shrink to a new steady state level length (Conrad *et al.* 1990; Lustig *et al.* 1990). This is a reversible effect, and on returning to 23°C the telomeres grow until they resume their wild-type length. These results suggest that RAP1 may have a role in telomere maintenance, as altering the amount of active RAP1 by mutation leads to altered telomere length. One explanation is that these temperature-sensitive mutations disrupt the non-nucleosomal telomeric chromatin structure associated with RAP1 binding (see below) and thus alter the accessibility of the telomeric DNA to proteins involved in telomere sequence addition.

There is genetic evidence that RAP1 is not the only protein present at yeast telomeres. In a cell also expressing wild-type RAP1, over-production of the C-terminus of RAP1, which cannot bind DNA, results in altered telomere length (Conrad *et al*. 1990). The most straightforward explanation is that this additional, N-terminally deleted version of RAP1 is titrating out another protein that normally binds wild-type RAP1 at telomeres. A sophisticated genetic screen has identified one such RAP1-interacting factor, RIF1 (Hardy *et al*. 1992). Strains carrying deletions of the *RIF1* gene grow normally but are defective in silencing and show increased telomere length. It is speculated that telomere elongation can result from reducing the amount of RIF1 bound at telomeres, either by titrating it out with a mutant RAP1 protein or by mutations in the *RIF1* gene. It remains to be determined if RIF1 binds DNA or whether it is a component of telomeric chromatin because of a protein–protein interaction with RAP1.

The interaction between RIF1 and RAP1 may explain the effect of certain *rap1* mutations on telomere length. For example, Sussel and Shore (1991) have isolated a series of viable *rap1^s* mutations which are defective in silencer function. These strains show no apparent growth defect, suggesting that the expression of the large number of essential genes whose promoters contain RAP1 binding sites is normal. However, *rap1^s* strains have longer telomeres. When sequenced, the mutations all cause amino acid changes in a very small region of the C-terminus, suggesting this region of the protein is involved in telomere length regulation. *rap1^s* mutations have a phenotype strikingly similar to a *rif1* mutation, and indeed *rap1^s* mutations appear to interfere with the RAP1–RIF1 interaction as determined using the two-hybrid system (Hardy *et al*. 1992). Similarly, intragenic suppressors of a *rap1^{ts}* mutation have been identified that allow such strains to grow at 37°C (Kyrion *et al*. 1992). They carry additional mutations in the *RAP1* gene, already carrying the *rap1^{ts}* mutations. These *rap1^t* mutations result in a very marked increase in telomere length and heterogeneity. These suppressor mutations result in C-terminal truncations of the RAP1 protein which retain DNA binding activity but would be predicted to lose their ability to interact with RIF1 (Henry *et al*. 1990; Hardy *et al*. 1992; Kyrion *et al*. 1992). However, the telomere length increase of strains carrying *rap1^t* mutations is much more dramatic than that for *rif1* strains, indicating that the telomere length phenotype is not solely the result of abolishing the RAP1–RIF1 interaction.

In summary, RAP1 is implicated in having an effect on telomere length regulation because *rap1^{ts}* mutations result in shortened telomeres. A second protein, RIF1, may also be a component of telomeric chromatin, perhaps via a protein–protein interaction with RAP1. RIF1 is implicated because telomere length is increased by mutations that reduce the amount of RIF1 complexed with RAP1 at telomeres, either by deleting the *RIF1* gene directly or by mutations such as *rap1ˢ* which are suggested to disrupt the RAP1–RIF1 association.

RAP1 is a very abundant protein ($>4 \times 10^3$ molecules/cell; Buchman *et al.* 1988*a*) and appears to be present in telomeric chromatin (Conrad *et al.* 1990). It is a component of the yeast nuclear scaffold (Cardenas *et al.* 1990) which is of interest as human telomeres are bound to the nuclear matrix (de Lange 1992). RAP1 can also mediate loop formation *in vitro* between adjacent RAP1 sites in a DNA molecule (Hofmann *et al.* 1989). Antibody studies have shown that the majority of the chromosome-bound RAP1 is localized to the ends of meiotic yeast chromosomes (Klein *et al.* 1992; Gilson *et al.* 1993*a*). Meiotic chromosomes were used in these experiments as mitotic chromosomes do not condense sufficiently to be visualized in budding yeast. In interphase the majority of the RAP1 signal is seen as approximately eight patches close to the nuclear periphery, consistent with the hypothesis that yeast telomeres are associated with each other and with the nuclear envelope. The number of foci is increased by mutations in *SIR3* and *SIR4*, which are required for transcriptional silencing at both telomeres and the silent mating-type loci (Palladino *et al.* 1993; see Chapter 8).

RAP1 binds duplex DNA; there is no evidence that it can bind a single-stranded terminus or that it has a preference for the terminus of double-stranded DNA. Binding to a single site *in vitro* can induce 90° bends in the DNA, which may have significant implications for chromatin structure (Vignais and Sentenac 1989; Gilson *et al.* 1993*b*). The mouse major satellite, which forms the pericentromeric heterochromatin, is also intrinsically bent (Radic *et al.* 1987). It is therefore tempting to speculate that altered DNA conformation may have a role in heterochromatin behaviour. Numerous RAP1 molecules can bind at the same time to naked telomeric DNA *in vitro*, creating long stretches (up to *c*. 100 bp) totally protected from DNase I digestion (Gilson *et al.* 1993*b*). RAP1 may therefore be the major double-stranded DNA binding protein at yeast telomeres. The TG_{1-3} component of yeast telomeres are found in a non-nucleosomal form and RAP1 appears to be one protein component of this 'telosome' (Wright *et al.* 1992).

The array of RAP1 binding sites at yeast telomeres appears to assemble an altered chromatin conformation which is associated with the transcriptional repression of telomere-adjacent genes (see Chapter 8). This repression involves proteins such as those encoded by the *SIR* genes which also act to establish repressive chromatin at the silent mating-type loci. This altered chromatin conformation has been shown to affect DNA–protein interactions *in vivo*, such as decreasing the accessibility of the telomere-adjacent DNA to nucleases and methylases (see Chapter 8). One possibility could be that this repressive chromatin influences the accessibility of proteins involved in telomere sequence addition to the TG_{1-3} tract and thus affects telomere length. There are a number of lines of evidence to suggest that this is not the case. The repressive chromatin requires the products of the *SIR* genes (Chapter 8) which may function through deacetylation of histone H4. For this *SIR*-dependent repressive chromatin to influence accessibility to the terminal TG_{1-3} tract would presumably require this altered chromatin to extend into the TG_{1-3} tract. This does not, however, appear to be the case. In fact it has been shown that the terminal TG_{1-3} tracts are assembled into a non-nucleosomal form, possibly because they are coated with RAP1 molecules (Wright *et al.* 1992). Although the TG_{1-3}-adjacent region, which is in assembled into nucleosomes, is protected from *dam* methylation in a *SIR*-dependent fashion (Gottschling 1992), there is no evidence that the TG_{1-3} tracts show a similar *SIR*-dependent protection. Instead accessibility to the TG_{1-3} sequences may depend on direct protection by the RAP1 protein itself. *In vitro* RAP1 alone can protect large stretches of TG_{1-3} DNA from being accessible to nucleases (Gilson *et al.* 1993*b*). Genetic evidence in support of this model is that mutations in *SIR3* and *SIR4* have little effect on telomere length despite relieving position effect repression of telomere-adjacent genes (Palladino *et al.* 1993). In contrast, *RAP1* mutations can have marked effects on telomere length. For example, *rap1ᵗ* mutations have a very dramatic effect on telomere length, which ranges from the wild-type 300 bp to over 4 kb (Kyrion *et al.* 1992). Truncation of the RAP1 protein by a *rap1ᵗ* mutation may make the TG_{1-3} DNA more accessible to proteins involved in telomere sequence addition. In summary it is possible to speculate that telomere length may be determined by accessibility of proteins to the TG_{1-3} arrays, and furthermore that this may be primarily a direct effect of bound RAP1 molecules, rather than a consequence of the repressive chromatin found in telomere-adjacent regions.

Other proteins

Liu and Tye (1991) have biochemically identified two additional proteins that bind yeast telomeres *in vitro*. TBFα binds to double-stranded DNA corresponding to yeast, *Tetrahymena*, and human telomere repeat sequences, whereas TBFβ only binds yeast TG_{1-3} sequences. TBFβ appears distinct from RAP1 as it does not bind a non-telomeric RAP1 site such as that found upstream of the *TEF2* gene.

TBFα footprints the junction between the terminal TG_{1-3} repeats and the subtelomeric X sequences found at some yeast telomeres (see Chapter 3). The two footprints obtained both have within them the sequence TTAGGG, and indeed TBFα will footprint a $(TTAGGG)_n$ array. It is encoded by the *TBF1* gene (Brigati *et al.* 1993), originally isolated by screening a λgt11 expression library with a labelled probe containing fifty TTAGGG repeats in double-stranded form. Disruption of the *TBF1* gene is lethal, but growth of a *tbf1^{ts}* allele at semi-permissive temperature does not result in altered telomere length (Brigati *et al.* 1993). It is unknown if TBFα and TBFβ interact with telomeres *in vivo*. Although RAP1 appears to be the main DNA binding protein for the TG_{1-3} repeats, with a possible role in forming the non-nucleosomal telomeric chromatin structure this does not exclude other proteins such as TBFα and β also having a role in telomere function.

Physarum polycephalum

This species has the same terminal repeat sequence as humans, $(TTAGGG)_n$. An unusually small (10 kDa) heat-stable protein termed PPT has been purified that binds *Physarum* telomeric sequences *in vitro*. It binds specifically to $(TTAGGG)_n$ in both single-and double-stranded form, with multiple repeats bound *in vitro*. There is 10-fold more protein in the cell than TTAGGG repeats, suggesting it may 'coat' the entire length of the telomere (Coren *et al.* 1991; Coren and Vogt 1992). *Physarum* telomeres do not have regularly spaced nucleosomes, which could be the result of PPT binding, but it has yet to be determined whether PPT binds *in vivo* (Lucchini *et al.* 1987).

Vertebrates

McKay and Cooke (1992*a*) have identified an abundant (*c.* 10^7 molecules per cell) protein in mouse nuclei that binds with some sequence-specificity to single-stranded $(TTAGGG)_n$ *in vitro*. Further analysis has identified this protein as heterogeneous ribonucleoprotein (hnRNP) A2/B1 (McKay and Cooke 1992*b*). Antibodies against this protein do not label the telomeres of mitotic chromosomes and give an intense uniform nuclear staining in interphase. A HeLa protein fraction that binds single-stranded $(TTAGGG)_n$ also contains hnRNPs (Ishikawa *et al.* 1993). These proteins bind RNA containing r(UUAGGG) repeats better than DNA, and with a high degree of sequence specificity (McKay and Cooke 1992*b*; Ishikawa *et al.* 1993). One *in vivo* role may be to bind the pre-mRNA 3′ splice site, whose consensus is $(Py)_nNPyAGG$ (Ishikawa *et al.* 1993). The involvement of hnRNPs with primary RNA transcripts in spliceosomes might make it seem unlikely that they are involved in telomere biology. However, RAP1 provides a precedent for an abundant telomere binding protein with additional functions related to transcription.

Zhong *et al.* (1992) have used a band-shift assay to identify a protein which binds double-stranded TTAGGG repeats. The activity is found in human, mouse, and monkey cells, with an apparent molecular mass of 50 kDa, but has yet to be purified to homogeneity. It binds TTAGGG repeats but not a variety of related sequences, nor does it bind single-stranded DNA.

Human telomeres are assembled into nucleosomal chromatin (Makarov *et al.* 1993) and are attached to the nuclear matrix (de Lange 1992). Terminal (TTAGGG)$_n$ arrays are attached, whereas internal (TTAGGG)$_n$ arrays are not. A factor from *Xenopus* eggs will specifically bind (TTAGGG)$_2$ when the sequence terminates a single-stranded 3′ overhang (Cardenas *et al.* 1993). However, the factor does not bind this sequence when it is at an internal location, when it terminates a 5′ protruding end, nor when it is present as double-stranded DNA. These observations suggest it is an end-binding factor, similar to the ciliate telomere binding proteins. It does not seem to be an hnRNP as it is found at low abundance (< 100 copies per somatic nucleus) and does not bind r(UUAGGG) repeats.

One set of proteins that might be expected to have a role in telomere behaviour are the nuclear lamins, which lie immediately underneath the nuclear envelope. In some species telomeres are associated with the nuclear envelope (see Chapter 2). However, in a filter-binding assay mouse lamins A and C bind oligonucleotides terminating in single-stranded TTAGGG with low affinity (*c.* 3×10^6 M^{-1}), a figure that is only three-fold higher than for their interaction with non-telomeric oligonucleotides (Shoeman and Traub 1990). Although these proteins clearly have some DNA binding activity that might be involved in the general attachment of chromatin to the nuclear envelope, the lack of a highly specific interaction between telomeres and nuclear lamins argues against the latter promoting telomere interactions with the nuclear envelope. Indeed, lamins A and C do not appear to have a preference for the telomeric regions of metaphase chromosomes *in vitro* but bind all along their length (Glass and Gerace 1990). As speculated in Chapter 2, these apparent specific interactions with the nuclear envelope could reflect telomeres being peripherally located at telophase and this organization becoming fixed in interphase by interactions with the nuclear matrix and lamina as the nucleus reforms around the chromosome group.

Some vertebrate proteins have been identified that will bind structures containing G-tetrads *in vitro*, and perhaps because of this will bind with some affinity to telomeric oligonucleotides in G-tetrad form (see Chapter 3). Although only speculation, it is possible that their specificity for certain sequences may be based more on the ability of the oligonucleotide to fold into a G-tetrad structure rather than being bound because of its primary DNA sequence. Many are unlikely to be authentic telomere binding proteins *in vivo*; for example, one such protein is the muscle-specific transcription factor MyoD. However, although they may not have a role at telomeres, their existence provides some evidence in support of the suggestion that G-tetrad structures have an *in vivo* function, even if not at telomeres.

Telomerase

Telomerase activity has been identified in a variety of cells including those of a number of ciliate species, humans and mice (see Chapter 4). *Euplotes* telomerase activity is eluted in a hypotonic wash of macronuclei, and purifies away from the telomere binding activity (Shippen-Lentz and Blackburn 1989). This argues that telomerase is not one of the identified ciliate telomere binding proteins.

It is of interest to investigate how telomere binding proteins interact with telomerase in the *in vitro* assays, which use single-stranded purified DNA as the substrate, because telomeric DNA complexed with its binding proteins may more closely resemble the *in vivo* substrate for telomerase. Although DNA bound by the *Oxytricha* telomere binding protein is protected from micrococcal nuclease digestion, it serves as an efficient substrate for *Oxytricha* telomerase as well as reverse transcriptase and Klenow DNA polymerase (Shippen *et al.* 1994). This indicates that although the phosphate backbone is protected from nuclease attack, the DNA bases themselves are exposed for replication.

Genetic screens for *trans*-acting factors in *S. cerevisiae*

Although telomere binding proteins might be identified genetically, one major problem is the nature of the phenotype; abolishing essential telomere function will eventually be lethal, for example by incomplete replication leading to the loss of essential subtelomeric genes or because breakage–fusion–bridge cycles occur (Chapter 2). With this in mind many genetic screens have centred on more subtle effects on telomere behaviour, in particular changes in telomere length. Indeed, it was the observation that telomere length varies among laboratory yeast strains that suggested *trans*-acting factors involved with telomeres might be identified by genetic methods (Walmsley and Petes 1985). With such approaches one must consider the potential rate of turnover of telomeric sequences, as this has important consequences for expression of the desired phenotype. For example, if one were to completely abolish telomere sequence addition, how fast would sequences be lost from the end? In *Drosophila*, chromosomes with broken ends recede at a rate which corresponds to only a few base-pairs per cell division (Chapter 9). In contrast, human primary fibroblasts in culture, which do not show telomerase activity, have telomeres that shorten at around 50 bp per division (Chapter 7). Therefore the rate of sequence turnover is such that a number of cell divisions may have to occur before a substantial change in telomere length is observed.

A number of mutations affecting telomere length in yeast have indeed been identified, using Southern analysis to screen strains for altered telomere length (Lustig and Petes 1986). Recessive mutations were found that fall into two complementation groups, *tel1* and *tel2*, and result in reduced telomere length. The new length is only achieved after a large number of generations (>100). A similar phenotypic lag is seen for *cdc17* (DNA polymerase α), *est1*, and *rap1* mutations, and may reflect the rate of turnover of telomeric sequences. The products of the *TEL1* and *TEL2* genes have yet to be identified and there is little known of their role at telomeres. Mutations in the *PIF1* gene, a 5' to 3' DNA helicase originally identified by its role in mitochondrial DNA stability, cause a modest increase in telomere length (Schulz and Zakian 1994). How many of these mutations identify proteins involved in telomere function, as opposed to producing indirect pleiotropic effects, remains unclear.

Runge and Zakian (1993) reasoned that a mutation that adversely affects telomere function would lead to increased loss of linear chromosomes. Some of the *lcs* mutations they identified cause an increased rate of loss of linear yeast chromosomes while not affecting circular chromosomes. Although the magnitude of the chromosome stability phenotype is not great (for example, a linear chromosome III in a diploid is correctly segregated in 97.8% of cell divisions in the most extreme *lcs* mutant, whereas in a wild-type cell the rate is 99.9%) this is to be expected; a mutation that severely affected telomere function would be lethal because all the chromosomes of the yeast host are linear. As discussed in Chapter 2, if the chromosome stabilization aspect of telomere function was completely abolished, so the chromosome end now behaved like a non-telomeric double-strand break, a very high rate of chromosome loss and lethality would be expected. No change in telomere length was seen in any of the *lcs* mutants, and in the absence of further evidence it remains unproven that the increased loss rate of linear chromosomes in any of these strains is caused by defects in telomere function.

est1

When a circular plasmid carrying two *Tetrahymena* telomeres separated by the *URA3* gene is introduced into yeast it resolves to a linear form at a low but detectable frequency. Linearization results in loss of the *URA3* gene and the cell becomes resistant to 5-fluoro-orotic acid (5-FOA). As the *Tetrahymena* telomeres are healed by the acquisition of yeast TG_{1-3} repeats, defects in telomere metabolism might affect this resolution reaction. This approach led to the isolation of the *est1* mutation.

est1 (ever shorter telomeres) is a mutation that results in a reduced rate of linearizing such a plasmid (Lundblad and Szostak 1989). Telomere length in *est1* mutants is altered, but it is not the progressive attainment of a new steady-state length seen with mutations such as *tel1*. Instead *est1* mutants inexorably lose telomeric TG_{1-3} repeats at the rate of a few base-pairs per division. Furthermore, the cultures appear to senesce, in that they progressively accumulate more and more dead cells (Lundblad and Szostak 1989). One interpretation is that telomere length maintenance has been disrupted, with telomeres getting progressively shorter until all the terminal repeat sequences are lost. However, the exact cause of cell death in these strains remains somewhat unclear.

The *est1* mutations do not have a tight phenotype; although the cultures do senesce, rapidly growing survivors appear at the later stages and take over the culture. One might have expected these to be simply revertants of the *est1* mutation, but as an *est1*Δ deletion strain also produces survivors this cannot be the case. Further analysis of these rapidly growing *est1*Δ survivors has revealed that a new pathway of telomere maintenance, involving amplification of the subtelomeric Y' elements, has become active in these cells (Lundblad and Blackburn 1993). The basis of this telomere maintenance pathway is described in more detail in Chapter 9.

Although telomerase activity has yet to be detected in yeast, the phenotype of the *est1* mutation is what one might predict for loss of telomerase activity. The protein predicted from the *EST1* sequence shows little sequence similarity to anything in the current databases. Sequence motifs found in a range of RNA-dependent DNA polymerases and RNA-dependent RNA polymerases can be found in the EST1 protein, but the match is too weak to prove that EST1 is an RNA-dependent polymerase; for example, many non-polymerase sequences match the motif better than EST1 (Lundblad and Blackburn 1990; Henikoff 1991). However, telomerase could be described as a specialized RNA-dependent DNA polymerase carrying its own template. It remains to be seen if EST1 is homologous to any telomerase proteins, as these have yet to be sequenced from any species.

cdc8, *cdc17*, and *cdc21*

CDC17 encodes the catalytic subunit of yeast DNA polymerase α, which is responsible for lagging strand chromosomal replication (Wang 1991). Strains carrying a *cdc17* mutation do not grow at 37°C. At a semi-permissive temperature, where growth rate is reduced due to limiting active polymerase, telomere length increases (Carson and Hartwell 1985). *cdc8* and *cdc21* mutations also affect telomere length, and the corresponding genes encode dTMP kinase and synthase respectively (cited in Lundblad and Szostak 1989). It is not known why these mutations affect telomere length but it is possible to speculate. For example, in the strand-slippage mechanism for telomere sequence addition (see Chapter 1) it is hypothesized that short Okazaki fragments produced by DNA polymerase α slip on repetitive templates, resulting in more sequence being produced than templated by the complementary strand. Perhaps then a *cdc17* mutation, by affecting DNA polymerase α, slows down lagging strand replication so that these short fragments are more long-lived, thus allowing additional time for slippage on telomeric templates and therefore an increase in telomere length. Similarly, *cdc8* and *cdc21* mutations are likely to affect the nucleotide pool and in addition to potentially affecting DNA polymerase activity this might also lead to a change in the activity of another DNA polymerase, namely telomerase.

Genetic screens in other species

There has been considerable success in identifying proteins involved in telomere biology in yeast, reflecting the powerful genetics available in this species. In both plants and mammals genotypic effects on telomere length have been observed, which suggests loci encoding proteins which interact with telomeres. However, it is unlikely that these proteins will be isolated by a genetic route given the difficulties in going from phenotype to gene in these species. There is also a major problem with phenotypic lag. For example, in mice, inbred strains can show differing ranges of telomere lengths, much greater than the difference between individuals of the same strain (Kipling and Cooke 1990; Starling *et al.* 1990). Although this is consistent with the existence of loci encoding proteins affecting telomere length in the mouse there is a phenotypic lag; the length of a telomere in an offspring is very similar to that in the parent, and it seems likely that many mouse generations have to occur before there is a significant percentage change in telomere length. Thus it would be very difficult in practice to use this as a phenotype with which to identify the corresponding genes.

Loci affecting telomere length have been identified in plants (reviewed by Chasan 1992). Different inbred maize lines show large differences in size of the $(TTTAGGG)_n$ smear on Southern blots, indicating differences in telomere length between the strains. However, all individuals of the same genotype have very similar telomere lengths, suggesting that telomere length is controlled by the genotype of the strain. Burr *et al.* (1992) have analysed telomere length in recombinant inbred (RI) mapping strains (see Burr and Burr 1991). There is a range of average telomere lengths in the various RI strains, with some in fact having shorter telomeres than either parent. This suggests the presence of multiple loci affecting telomere length, both positive and negative in influence. In a similar fashion there are strain-specific variations in telomere length in *Arabidopsis thaliana*. The two common ecotypes, Columbia and Landsberg, show reproducible differences in the lengths of their terminal repeat arrays. For example, one particular Landsberg telomere, detected with a unique subtelomeric probe, is about 600 bp longer than the equivalent one in Columbia (Richards *et al.* 1992). The terminal restriction maps appear identical in every other way, and although a subtle change cannot be ruled out, the difference is most simply explained by differences in the number of terminal $(TTTAGGG)_n$ repeats.

What determines telomere length?

The question of how cells regulate telomere length does not have the simple answer one might expect. A situation with a balance between the processes of telomere addition (e.g. telomerase) and loss (e.g. the end-replication problem) is not stable over a long period of time unless perfectly balanced. If one rate is even slightly greater than the other the result is inexorable growth or shrinkage. For telomere length to remain stable there must be some mechanism whereby telomere length feeds back to the processes of loss or addition and adjusts them accordingly.

Oxytricha provides one example of the problem. The macronuclear telomeres are very homogeneous, with both strands terminating at a precise position:

```
NNNNNNNNGGGGTTTTGGGGTTTTGGGGTTTTGGGGTTTTGGGGTTTTGGGG-3'
NNNNNNNNCCCCAAAACCCCAAAACCCCAAAACCCC-5'
```

In a growing and replicating macronucleus how is the fixed number of repeats maintained, especially considering that telomerase is processive *in vitro*, adding many repeats to a terminus? There is also a problem with the C-strand, which is replicated by RNA primase and DNA polymerase α. How does it manage to commence DNA synthesis exactly 17 nucleotides from the 3' terminus? In fact it may not; one possibility is that there is unregulated sequence addition on both strands, with a sequence-specific endonuclease cleaving subsequently, releasing the correct structure (Vermeesch *et al.* 1993; Vermeesch and Price 1994). These questions may be explained with a better understanding of the biochemistry of telomerase and telomere binding proteins.

Telomere length can be dynamic and can vary according to the growth of the cell. For example, in a growing culture of trypanosomes terminal restriction fragments lengthen with time (Bernards *et al.* 1983; Pays *et al.* 1983). Similarly, log phase *Tetrahymena* cultures show telomere growth to a new steady-state level (Larson *et al.* 1987) although often the cultures are taken over by more rapidly dividing cells with short, non-lengthening telomeres. Telomeres shorten in stationary phase *Tetrahymena* cultures, even though the cells are not dividing. In contrast, *S. cerevisiae* telomeres lengthen in stationary phase cultures and actually shrink slightly during log phase growth (Shampay and Blackburn 1988). This suggests that the presence of telomere sequences may be more important than their length.

What does the cell measure as its criteria for 'telomere length'? One possibility is the total amount of terminal repeat sequence in the cell, for example by titrating a sequence-specific DNA binding factor. One argument against this is that telomere length is remarkably stable in yeast in response to the introduction of extra telomeric sequences. Introduction of a massive excess of yeast telomeres or telomeric sequences on replicating circular or linear plasmids results in only a mild change in average telomere length. For example, a 25-fold excess of telomeres gives only a 50% increase in telomere length (Runge and Zakian 1989). Furthermore, there is strain-specificity as to whether extra TG_{1-3} sequence results in telomere lengthening or shortening (Constable *et al.* 1990). This suggests that this simple model of telomere length regulation does not function in yeast. Another argument against this model is that by itself it would be unable to prevent length fluctuations at any individual telomere, which might occasionally lead to a complete loss of all its terminal repeats. This would presumably be a particular problem for species with a large number of chromosomes. It would also have to be exquisitely sensitive to exert control in species where there are large internal blocks of telomere repeat sequence (see Chapter 3) and where loss of all the terminal repeats would result in only a small change in the total amount of telomere repeat sequence. This argues against a simple model where telomere length is determined simply on the basis of the total amount of terminal repeat sequence in the nucleus.

Another possibility is that cells are able to monitor and regulate the length of individual telomeres. For example, a factor bound to a sub-telomeric sequence could influence the access of other proteins to the very end of the chromosome. Physical interactions between proteins could make an end closer to the 'reference point' a better substrate for telomerase or a protein that protects the end of the chromosome from nuclease attack, and thus a shorter telomere would tend to lengthen. However, the telomeres of YACs are stable, arguing that no sequence other than the TG_{1-3} array is necessary to enable telomere length to be regulated. It therefore remains unclear how telomere length is regulated.

Summary

Telomere binding proteins are likely to play a role in protecting the ends of a chromosome from end-to-end fusion, attack from nucleases, and activating cell cycle checkpoints. They may also confer an atypical chromatin conformation on telomeric sequences, although what role this might play is less clear. The best characterized yeast telomere binding protein, RAP1, has additional well-defined roles as a transcriptional regulator at non-telomeric loci. The precedent of its multiple functions illustrates the caution required before dismissing any unexpected telomere binding activity as irrelevant. Further biochemical analysis of telomere binding proteins may well shed light on questions such as the mechanism of telomere length regulation and the control of telomerase activity *in vivo*.

References

Reviews

Blackburn, E. H. (1991). Structure and function of telomeres. *Nature*, **350**, 569–73.
Burr, B. and Burr, F. A. (1991). Recombinant inbreds for molecular mapping in maize: theoretical and practical considerations. *Trends Genet*, **7**, 55–60.
Chasan, R. (1992). Maize telomeres – the end of the line. *Plant Cell*, **4**, 865–7.

Gilson, E., Laroche, T., and Gasser, S. M. (1993a). Telomeres and the functional architecture of the nucleus. *Trends Cell Biol.*, **3**, 128–34.

Henderson, E. R. and Larson, D. D. (1991). Telomeres—what's new at the end? *Curr. Opin. Genet. Dev.*, **1**, 538–43.

Jackson, D. A. (1991). Structure-function relationships in eukaryotic nuclei. *Bioessays*, **13**, 1–10.

Wang, T. S.-F. (1991). Eukaryotic DNA polymerases. *Annu. Rev. Biochem.*, **60**, 513–52.

Zakian, V. A. (1989). Structure and function of telomeres. *Annu. Rev. Genet.*, **23**, 579–604.

Primary papers

Berman, J., Tachibana, C. Y., and Tye, B.-K. (1986). Identification of a telomere-binding activity from yeast. *Proc. Natl. Acad. Sci. USA*, **83**, 3713–17.

Bernards, A., Michels, P. A. M., Lincke, C. R., and Borst, P. (1983). Growth of chromosome ends in multiplying trypanosomes. *Nature*, **303**, 592–7.

Blackburn, E. H. and Chiou, S.-S. (1981). Non-nucleosomal packaging of a tandemly repeated DNA sequence at termini of extrachromosomal DNA coding for rRNA in *Tetrahymena*. *Proc. Natl. Acad. Sci. USA*, **78**, 2263–7.

Brigati, C., Kurtz, S., Balderes, D., Vidali, G., and Shore, D. (1993). An essential yeast gene encoding a TTAGGG repeat-binding protein. *Mol. Cell. Biol.*, **13**, 1306–14.

Buchman, A. R., Lue, N. F., and Kornberg, R. D. (1988a). Connections between transcriptional activators, silencers, and telomeres as revealed by functional analysis of a yeast DNA-binding protein. *Mol. Cell. Biol.*, **8**, 5086–99.

Buchman, A. R., Kimmerly, W. J., Rine, J., and Kornberg, R. D. (1988b). Two DNA binding factors recognize specific sequences at silencers, upstream activating sequences, autonomously replicating sequences, and telomeres in *Saccharomyces cerevisiae*. *Mol. Cell. Biol.*, **8**, 210–25.

Budarf, M. L. and Blackburn, E. H. (1986). Chromatin structure of the telomeric region and 3'-nontranscribed spacer of *Tetrahymena* ribosomal RNA genes. *J. Biol. Chem.*, **261**, 363–9.

Burr, B., Burr, F. A., Matz, E. C., and Romero-Severson, J. (1992). Pinning down loose ends: mapping telomeres and factors affecting their length. *Plant Cell*, **4**, 953–60.

Cardenas, M. E., Laroche, T., and Gasser, S. M. (1990). The composition and morphology of yeast nuclear scaffolds. *J. Cell Sci.*, **96**, 439–50.

Cardenas, M. E., Bianchi, A., and de Lange, T. (1993). A *Xenopus* egg factor with DNA-binding properties characteristic of terminus-specific telomeric proteins. *Genes Dev.*, **7**, 883–94.

Carson, M. J. and Hartwell, L. (1985). *CDC17*: an essential gene that prevents telomere elongation in yeast. *Cell*, **42**, 249–57.

Chambers, A., Tsang, J. S. H., Stanway, C., Kingsman, A. J., and Kingsman, S.M. (1989). Transcriptional control of the *Saccharomyces cerevisiae PGK* gene by RAP1. *Mol. Cell. Biol.*, **9**, 5516–24.

Constable, A., Feipeng, L., and Walmsley, R. M. (1990). Yeast telomere length varies in response to changes in the amount of polyC$_{1-3}$A in the cell. *Mol. Gen. Genet.*, **221**, 280–82.

Conrad, M. N., Wright, J. H., Wolf, A. J., and Zakian, V. A. (1990). RAP1 protein interacts with yeast telomeres *in vivo*: overproduction alters telomere structure and decreases chromosome stability. *Cell*, **63**, 739–50.

Coren, J. S. and Vogt, V. M. (1992). Purification of a telomere-binding protein from *Physarum polycephalum. Biochim. Biophys. Acta*, **1171**, 162–6.

Coren, J. S., Epstein, E. M., and Vogt, V. M. (1991). Characterization of a telomere-binding protein from *Physarum polycephalum. Mol. Cell. Biol.*, **11**, 2282–90.

de Lange, T. (1992). Human telomeres are attached to the nuclear matrix. *EMBO J.*, **11**, 717–24.

Edwards, C. A. and Firtel, R. A. (1984). Site-specific phasing in the chromatin of the rDNA in *Dictyostelium discoideum. J. Mol. Biol.*, **180**, 73–90.

Fang, G. and Cech, T. R. (1991). Molecular cloning of telomere-binding protein genes from *Stylonychia mytilis. Nucleic Acids Res.*, **19**, 5515–18.

Fang, G. and Cech, T. R. (1993a). *Oxytricha* telomere-binding protein: DNA-dependent dimerization of the α and β subunits. *Proc. Natl. Acad. Sci. USA*, **90**, 6056–60.

Fang, G. and Cech, T. R. (1993b). The β subunit of Oxytricha telomere-binding protein promotes G-quartet formation of telomeric DNA. *Cell*, **74**, 875–85.

Fang, G., Gray, J. T., and Cech, T. R. (1993). *Oxytricha* telomere-binding protein: separable DNA-binding and dimerization domains of the α-subunit. *Genes Dev.*, **7**, 870–82.

Gilson, E., Roberge, M., Giraldo, R., Rhodes, D., and Gasser, S. M. (1993b). Distortion of the DNA double helix by RAP1 at silencers and multiple telomeric binding sites. *J. Mol. Biol.*, **231**, 293–310.

Glass, J. R. and Gerace, L. (1990). Lamins A and C bind and assemble at the surface of mitotic chromosomes. *J. Cell Biol.*, **111**, 1047–57.

Gottschling, D. E. (1992). Telomere-proximal DNA in *Saccharomyces cerevisiae* is refractory to methyltransferase activity *in vivo. Proc. Natl. Acad. Sci. USA*, **89**, 4062–5.

Gottschling, D. E. and Zakian, V. A. (1986). Telomere proteins: specific recognition and protection of the natural termini of Oxytricha macronuclear DNA. *Cell*, **47**, 195–205.

Graham, I. R. and Chambers, A. (1994). Use of a selection technique to identify the diversity of binding sites for the yeast RAP1 transcription factor. *Nucleic Acids Res.*, **22**, 124–30.

Gray, J. T., Celander, D. W., Price, C. M., and Cech, T. R. (1991). Cloning and expression of genes for the Oxytricha telomere-binding protein: specific subunit interactions in the telomeric complex. *Cell*, **67**, 807–14.

Hardy, C. F. J., Sussel, L., and Shore, D. (1992). A RAP1-interacting protein involved in transcriptional silencing and telomere length regulation. *Genes Dev.*, **6**, 801–14.

Henikoff, S. (1991). Playing with blocks: some pitfalls of forcing multiple alignments. *New Biol.*, **3**, 1148–54.

Henry, Y. A. L., Chambers, A., Tsang, J. S. H., Kingsman, A. J., and Kingsman, S. M. (1990). Characterisation of the DNA binding domain of the yeast RAP1 protein. *Nucleic Acids Res.*, **18**, 2617–23.

Hicke, B. J., Celander, D. W., MacDonald, G. H., Price, C. M., and Cech, T. R. (1990). Two versions of the gene encoding the 41-kilodalton subunit of the telomere binding protein of *Oxytricha nova. Proc. Natl. Acad. Sci. USA*, **87**, 1481–5.

Hofmann, J. F.-X., Laroche, T., Brand, A. H., and Gasser, S. M. (1989). RAP-1 factor is necessary for DNA loop formation *in vitro* at the silent mating type locus *HML*. *Cell*, **57**, 725-37.

Huet, J. and Sentenac, A. (1987). TUF, the yeast DNA-binding factor specific for UAS_{rpg} upstream activating sequences: identification of the protein and its DNA-binding domain. *Proc. Natl. Acad. Sci. USA*, **84**, 3648-52.

Ishikawa, F., Matunis, M. J., Dreyfuss, G., and Cech, T. R. (1993). Nuclear proteins that bind the pre-mRNA 3' splice site sequence r(UUAG/G) and the human telomeric DNA sequence $d(TTAGGG)_n$. *Mol. Cell. Biol.*, **13**, 4301-10.

Kipling, D. and Cooke, H. J. (1990). Hypervariable ultra-long telomeres in mice. *Nature*, **347**, 400-2.

Klein, F., Laroche, T., Cardenas, M. E., Hofmann, J. F.-X., Schweizer, D., and Gasser, S. M. (1992). Localization of RAP1 and topoisomerase II in nuclei and meiotic chromosomes of yeast. *J. Cell Biol.*, **117**, 935-48.

Kyrion, G., Boakye, K. A., and Lustig, A. J. (1992). C-terminal truncation of RAP1 results in the deregulation of telomere size, stability, and function in *Saccharomyces cerevisiae*. *Mol. Cell. Biol.*, **12**, 5159-73.

Larson, D. D., Spangler, E. A., and Blackburn, E. H. (1987). Dynamics of telomere length variation in Tetrahymena thermophila. *Cell*, **50**, 477-83.

Lipps, H. J., Gruissem, W., and Prescott, D. M. (1982). Higher order DNA structure in macronuclear chromatin of the hypotrichous ciliate *Oxytricha nova*. *Proc. Natl. Acad. Sci. USA*, **79**, 2495-9.

Liu, Z. and Tye, B.-K. (1991). A yeast protein that binds to vertebrate telomeres and conserved yeast telomeric junctions. *Genes Dev.*, **5**, 49-59.

Lohman, T. M. and Overman, L. B. (1985). Two binding modes in *Escherichia coli* single strand binding protein–single stranded DNA complexes. Modulation by NaCl concentration. *J. Biol. Chem.*, **260**, 3594-603.

Longtine, M. S., Wilson, N. M., Petracek, M. E., and Berman, J. (1989). A yeast telomere binding activity binds to two related telomere sequence motifs and is indistinguishable from RAP1. *Curr. Genet.*, **16**, 225-39.

Longtine, M. S., Enomoto, S., Finstad, S. L., and Berman, J. (1993). Telomere-mediated plasmid segregation in *Saccharomyces cerevisiae* involves gene products required for transcriptional repression at silencers and telomeres. *Genetics*, **133**, 171-82.

Lucchini, R., Pauli, U., Braun, R., Koller, T., and Sogo, J. M. (1987). Structure of the extrachromosomal ribosomal RNA chromatin of *Physarum polycephalum*. *J. Mol. Biol.*, **196**, 829-43.

Lundblad, V. and Blackburn, E. H. (1990). RNA-dependent polymerase motifs in EST1: tentative identification of a protein component of an essential yeast telomerase. *Cell*, **60**, 529-30.

Lundblad, V. and Blackburn, E. H. (1993). An alternative pathway for yeast telomere maintenance rescues *est1*⁻ senescence. *Cell*, **73**, 347-60.

Lundblad, V. and Szostak, J. W. (1989). A mutant with a defect in telomere elongation leads to senescence in yeast. *Cell*, **57**, 633-43.

Lustig, A. J. and Petes, T. D. (1986). Identification of yeast mutants with altered telomere structure. *Proc. Natl. Acad. Sci. USA*, **83**, 1398-402.

Lustig, A. J., Kurtz, S., and Shore, D. (1990). Involvement of the silencer and UAS binding protein RAP1 in regulation of telomere length. *Science*, **250**, 549-53.

McKay, S. J. and Cooke, H. (1992a). A protein which specifically binds to single stranded $TTAGGG_n$ repeats. *Nucleic Acids Res.*, **20**, 1387-91.

McKay, S. J. and Cooke, H. (1992b). hnRNP A2/B1 binds specifically to single stranded vertebrate telomeric repeat TTAGGG$_n$. *Nucleic Acids Res.*, **20**, 6461–6464.

Makarov, V. L., Lejnine, S., Bedoyan, J., and Langmore, J. P. (1993). Nucleosomal organization of telomere-specific chromatin in rat. *Cell*, **73**, 775–87.

Meyer, G. F. and Lipps, H. J. (1981). The formation of polytene chromosomes during macronuclear development of the hypotrichous ciliate *Stylonychia mytilus*. *Chromosoma*, **82**, 309–14.

Moehle, C. M. and Hinnebusch, A. G. (1991). Association of RAP1 binding sites with stringent control of ribosomal protein gene transcription in *Saccharomyces cerevisiae*. *Mol. Cell. Biol.*, **11**, 2723–35.

Palladino, F., Laroche, T., Gilson, E., Axelrod, A., Pillus, L., and Gasser, S. M. (1993). SIR3 and SIR4 proteins are required for the positioning and integrity of yeast telomeres. *Cell*, **75**, 543–55.

Pays, E., Laurent, M., Delinte, K., Van Meirvenne, N., and Steinert, M. (1983). Differential size variations between transcriptionally active and inactive telomeres of *Trypanosoma brucei*. *Nucleic Acids Res.*, **11**, 8137–47.

Prescott, D. M. (1983). The C-value paradox and genes in ciliated protozoa. *Modern Cell Biol.*, **2**, 329–52.

Price, C. M. (1990). Telomere structure in *Euplotes crassus*: characterization of DNA-protein interactions and isolation of a telomere-binding protein. *Mol. Cell. Biol.*, **10**, 3421–31.

Price, C. M. and Cech, T. R. (1987). Telomeric DNA–protein interactions of *Oxytricha* macronuclear DNA. *Genes Dev.*, **1**, 783–93.

Price, C. M. and Cech, T. R. (1989). Properties of the telomeric DNA-binding protein from *Oxytricha nova*. *Biochemistry*, **28**, 769–74.

Price, C. M., Skopp, R., Krueger, J., and Williams, D. (1992). DNA recognition and binding by the *Euplotes* telomere protein. *Biochemistry*, **31**, 10835–43.

Radic, M. Z., Lundgren, K., and Hamkalo, B. A. (1987). Curvature of mouse satellite DNA and condensation of heterochromatin. *Cell*, **50**, 1101–8.

Raghuraman, M. K. and Cech, T. R. (1989). Assembly and self-association of Oxytricha telomeric nucleoprotein complexes. *Cell*, **59**, 719–28.

Raghuraman, M. K. and Cech, T. R. (1990). The effect of monovalent cation-induced telomeric DNA structure on the binding of *Oxytricha* telomeric protein. *Nucleic Acids Res.*, **18**, 4543–52.

Raghuraman, M. K., Dunn, C. J., Hicke, B. J., and Cech, T. R. (1989). *Oxytricha* telomeric nucleoprotein complexes reconstituted with synthetic DNA. *Nucleic Acids Res.*, **17**, 4235–53.

Richards, E. J., Chao, S., Vongs, A., and Yang, J. (1992). Characterization of *Arabidopsis thaliana*, telomeres isolated in yeast. *Nucleic Acids Res.*, **20**, 4039–46.

Runge, K. W. and Zakian, V. A. (1989). Introduction of extra telomeric DNA sequences into *Saccharomyces cerevisiae* results in telomere elongation. *Mol. Cell. Biol.*, **9**, 1488–97.

Runge, K. W. and Zakian, V. A. (1993). *Saccharomyces cerevisiae* linear chromosome stability (*lcs*) mutants increase the loss rate of artificial and natural linear chromosomes. *Chromosoma*, **102**, 207–17.

Schulz, V. P. and Zakian, V. A. (1994). The Saccharomyces *PIF1* DNA helicase inhibits telomere elongation and de novo telomere formation. *Cell*, **76**, 145–55.

Shampay, J. and Blackburn, E. H. (1988). Generation of telomere-length heterogeneity in *Saccharomyces cerevisiae*. *Proc. Natl. Acad. Sci. USA*, **85**, 534–8.

Shippen, D. E., Blackburn, E. H., and Price, C. M. (1994). DNA bound by the

Oxytricha telomere protein is accessible to telomerase and other DNA polymerases. *Proc. Natl. Acad. Sci. USA*, **91**, 405–9.

Shippen-Lentz, D. and Blackburn, E. H. (1989). Telomere terminal transferase activity from *Euplotes crassus* adds large numbers of TTTTGGGG repeats onto telomeric primers. *Mol. Cell. Biol.*, **9**, 2761–4.

Shoeman, R. L. and Traub, P. (1990). The *in vitro* DNA-binding properties of purified nuclear lamin proteins and vimentin. *J. Biol. Chem.*, **265**, 9055–61.

Shoeman, R. L., Wadle, S., Scherbarth, A., and Traub, P. (1988). The binding *in vitro* of the intermediate filament protein vimentin to synthetic oligonucleotides containing telomere sequences. *J. Biol. Chem.*, **263**, 18744–9.

Shore, D. and Nasmyth, K. (1987). Purification and cloning of a DNA binding protein from yeast that binds to both silencer and activator elements. *Cell*, **51**, 721–32.

Starling, J. A., Maule, J., Hastie, N. D., and Allshire, R. C. (1990). Extensive telomere repeat arrays in mouse are hypervariable. *Nucleic Acids Res.*, **18**, 6881–8.

Sussel, L. and Shore, D. (1991). Separation of transcriptional activation and silencing functions of the *RAP1*-encoded repressor/activator protein 1: isolation of viable mutants affecting both silencing and telomere length. *Proc. Natl. Acad. Sci. USA*, **88**, 7749–53.

Vermeesch, J. R. and Price, C. M. (1994). Telomeric DNA sequence and structure following *de novo* telomere synthesis in *Euplotes crassus*. *Mol. Cell. Biol.*, **14**, 554–66.

Vermeesch, J. R., Williams, D., and Price, C. M. (1993). Telomere processing in *Euplotes*. *Nucleic Acids Res.*, **21**, 5366–71.

Vignais, M.-L. and Sentenac, A. (1989). Asymmetric DNA bending induced by the yeast multifunctional factor TUF. *J. Biol. Chem.*, **264**, 8463–6.

Vignais, M.-L., Woudt, L. P., Wassenaar, G. M., Mager, W. H., Sentenac, A., and Planta, R. J. (1987). Specific binding of TUF factor to upstream activation sites of yeast ribosomal protein genes. *EMBO J.*, **6**, 1451–7.

Walmsley, R. M. and Petes, T. D. (1985). Genetic control of chromosome length in yeast. *Proc. Natl. Acad. Sci. USA*, **82**, 506–10.

Wang, W., Skopp, R., Scofield, M., and Price, C. (1992). *Euplotes crassus* has genes encoding telomere-binding proteins and telomere-binding protein homologs. *Nucleic Acids Res.*, **20**, 6621–9.

White, M. A., Wierdl, M., Detloff, P., and Petes, T. D. (1991). DNA-binding protein RAP1 stimulates meiotic recombination at the *HIS4* locus in yeast. *Proc. Natl. Acad. Sci. USA*, **88**, 9755–9.

Wright, J. H., Gottschling, D. E., and Zakian, V. A. (1992). *Saccharomyces* telomeres assume a non-nucleosomal chromatin structure. *Genes Dev.*, **6**, 197–210.

Zhong, Z., Shiue, L., Kaplan, S., and de Lange, T. (1992). A mammalian factor that binds telomeric TTAGGG repeats in vitro. *Mol. Cell. Biol.*, **12**, 4834–43.

6

Genome rearrangements and telomeres

What is the substrate for telomere addition *in vivo*? Much of what we know of this comes from analysis of the process of programmed genome fragmentation and *de novo* telomere addition in ciliates such as *Oxytricha* and *Tetrahymena* and the nematode *Ascaris*. The unusual life cycle of these organisms will be reviewed. The analysis of *de novo* telomere addition suggests that telomere sequence addition occurs onto A + T-rich sequences which would not be predicted to be telomerase substrates *in vitro*.

The molecular details of spontaneous breakage and healing events have been described in a number of species including humans and *Plasmodium*, where again they provide information on the *in vivo* substrate requirements for telomere sequence addition to compare with *in vitro* data on telomerase activity. Finally chromosome healing events in *S. cerevisiae* and to what extent they involve recombinational events will be reviewed.

Chromosome healing

A natural chromosome end behaves differently from a simple double-strand break, which is prone to recombination and fusion, and can activate cell cycle arrest. Double-strand breaks can occur in a number of ways; they can arise spontaneously, be the result of a programmed breakage process, or result from the introduction of a linear DNA molecule into the cell. Healing is the acquisition of telomere behaviour by such a DNA end, and is usually associated with the addition of telomere sequences, a number of examples of which have been described in Chapter 3. Often they involve the transfer of only terminal repeat sequences to the broken end, which suggests that the terminal repeats are sufficient for essential telomere function.

However, the healing of a broken chromosome end by *de novo* addition of telomeric sequences also enables another important question to be addressed: what is the substrate for telomere sequence addition *in vivo*? In particular, in those species where telomerase has been identified (*Oxytricha, Euplotes, Tetrahymena*, humans, and mice; see Chapter 4) it is possible to compare the *in vitro* substrates of telomerase with the *in vivo* sequences to which telomeric sequences are added. In these species there are numerous examples where terminal repeats have been added *in vivo* onto a sequence which has little sequence similarity to what would be expected to function

as a telomerase primer *in vitro*. Despite this, by expressing mutant telomerase RNA templates in *Tetrahymena* it has been shown that telomerase
is indeed responsible for this sequence addition. One interpretation is that
there may be additional factors *in vivo* which promote the recognition of
an end by telomerase.

Programmed genome rearrangements

Much of what is known regarding the healing of breaks *in vivo* comes from
the study of the programmed chromosome breakage and healing reactions
which occur during the development of ciliate macronuclei. Ciliates have
a very complex life cycle, the details of which vary from species to species
(reviewed by Prescott, 1992*a,b*). The most important feature with respect
to telomere biology is that during macronuclear development the genome
is fragmented and *de novo* telomere sequence addition occurs. By comparing genomic sequences before and after breakage the substrates for
in vivo telomere addition can be deduced. Another system which provides
an example of genome fragmentation and telomere addition is chromatin
diminution; this occurs in a wide range of species, although the molecular
details are perhaps best understood in nematodes such as *Ascaris*.

The ciliate life cycle

Ciliates are single-celled organisms which contain two kinds of nuclei: a
germ-line micronucleus and a vegetative macronucleus. The micronucleus
of *Oxytricha nova* contains high molecular weight DNA, is transcriptionally
silent, divides by conventional mitosis during vegetative growth, and undergoes meiosis during cell mating. In contrast the macronucleus contains
small, gene-sized DNA molecules and divides amitotically. It is transcriptionally active and is responsible for all gene expression during vegetative
growth.

The macronucleus is derived from a micronucleus by a series of programmed developmental events that occur after mating (Fig. 6.1). During
mating a cytoplasmic bridge forms between the two cells. The diploid
micronuclei in both cells undergo meiosis and the two cells exchange
haploid micronuclei. Two haploid micronuclei, one from each cell, fuse to
form a new diploid micronucleus, which then undergoes the series of events
shown in Fig. 6.1. At the same time the old macronucleus degenerates,
along with any remaining diploid or haploid micronuclei.

Development begins with multiple rounds of replication to form polytene
chromosomes. Following this much of the genome is eliminated. In *O. nova*
all the repetitive sequences (*c.* 40% of the micronuclear genome) and 95%
of the single-copy sequences (*c.* 60% of the micronuclear genome) are
eliminated. At least three different classes of sequence are eliminated. In

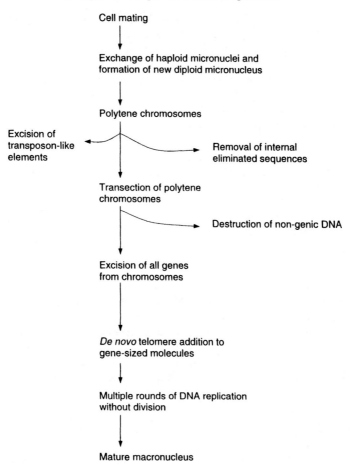

Cell mating

Exchange of haploid micronuclei and
formation of new diploid micronucleus

Polytene chromosomes

Excision of
transposon-like
elements

Removal of internal
eliminated sequences

Transection of polytene
chromosomes

Destruction of non-genic DNA

Excision of all genes
from chromosomes

De novo telomere addition to
gene-sized molecules

Multiple rounds of DNA replication
without division

Mature macronucleus

Fig. 6.1 Chromosome rearrangements during macronuclear development in hypotrichous ciliates. In parallel to the development of a macronucleus from a new diploid micronucleus after cell mating, the old macronucleus and any unused micronuclei are degraded. This sequence of events takes around 4 days.

some species (e.g. *Euplotes*) there is a massive excision and elimination of transposon-like elements. Secondly, all genes so far analysed are interrupted by short, internal eliminated sequences (IESs) which are spliced out to make the functional, uninterrupted macronuclear genes. Finally, there is massive elimination of other repetitive non-genic sequences. The excised sequences are destroyed by an unknown mechanism.

After IES and transposon excision the polytene chromosomes are cut up band by band and all the genes excised as small linear molecules. These

molecules have an average size of 2.2 kb in *Oxytricha*. This fragmentation is accompanied by the destruction of the non-genic sequences. All
Oxytricha gene-sized molecules studied so far contain a single transcription
unit and generally contain only a few hundred base pairs of non-coding
sequence. Each of these molecules is amplified to about 1000 copies in the
mature macronucleus, the result being a somatic nucleus which is highly
polyploid and very streamlined. This polyploidy is presumably an adapta-
tion to the size of the cell, which can be as large as 3 mm in length in some
ciliate species. Most importantly, genome fragmentation is followed by *de
novo* telomere sequence addition onto the new ends, and it is this process
which allows an *in vivo* analysis of the sequence requirements for telomere
addition.

Sequence analysis of the sites of breakage and healing

Oxytricha nova

This species contains about 2×10^7 gene-sized molecules. The region of
both the macronuclear and micronuclear genome corresponding to a
number of sites of breakage and telomere addition have been sequenced.
There are no pre-existing $(TTTTGGGG)_n$ repeats in the micronuclear
genome at the sites of telomere addition. The repeats are added *de novo*
onto A + T-rich sequences with no marked sequence similarity to each other
(Klobutcher *et al.* 1984, 1986, 1988). Similar results have been obtained in
O. fallax (Herrick *et al.* 1987).

Tetrahymena thermophila

The micronuclear genome consists of five pairs of chromosomes. During
macronuclear development around 15% of the genome is eliminated.
Elimination occurs at several thousand sites by deletion and rejoining. The
genome is also fragmented by breakage at between 50 and 200 specific sites.
The final result is a macronuclear genome consisting of about 200–300
chromosomal fragments which range in size from 20 kb to greater than
1000 kb. Breakage is directed by a 15 bp chromosome breakage sequence
(Cbs) motif which is found close to all the breakpoints and is present in
the region eliminated from the genome (Yao *et al.* 1987). Insertion of the
15 bp Cbs motif is sufficient to allow breakage at new sites in the genome
(Yao *et al.* 1990).

   Breakage is accompanied by *de novo* addition of $(TTGGGG)_n$ sequences
to the newly formed ends. A number of macronuclear telomeres and the
corresponding region of the micronuclear genome have been sequenced
(Yokoyama and Yao 1986; Yao *et al.* 1987; Austerberry and Yao 1988;
Spangler *et al.* 1988; Austerberry *et al.* 1989; Yu and Blackburn 1991).

In all the cases except the rDNA telomere (King and Yao 1982) there are no (TTGGGG)$_n$ repeats pre-existing in the micronuclear genome at or near the site of telomere addition. Furthermore, (TTGGGG)$_n$ is added onto very A+T-rich sequences which are unlike the G-rich sequences which function as primers for *Tetrahymena* telomerase *in vitro*. Indeed, *Tetrahymena* telomerase is unable to utilize an oligonucleotide corresponding to one such A+T-rich sequence as a primer *in vitro* (Spangler *et al.* 1988). However, there is direct evidence that telomerase is responsible for this sequence addition *in vivo* from experiments involving the expression of mutant telomerase template RNA molecules *in vivo* (detailed in Chapter 4). The marked difference in the apparent *in vivo* and *in vitro* substrates for *Tetrahymena* telomerase could result from additional factors which promote an interaction between telomerase and these A+T-rich sequences *in vivo*. Substrate choice by telomerase *in vitro* is discussed in detail in Chapter 4.

Paramecium

Various developmental breakpoints healed by telomere addition have been sequenced. As with other ciliates no telomeric sequences pre-exist at the breakpoint (Baroin *et al.* 1987; Forney and Blackburn 1988). There is also little if any sequence requirement for the healing of microinjected linear DNAs, which again suggests that pre-existing sequences are not necessary to promote telomere addition. For example, Gilley *et al.* (1988) have injected *Paramecium* with a pUC-like vector linearized by restriction enzyme digestion in the polylinker. This molecule consists entirely of plasmid vector sequences yet is maintained in the macronucleus at high copy number and has *Paramecium* (TT[T/G]GGG)$_n$ terminal repeat sequences added onto both ends.

Chromatin diminution in *Ascaris*

Programmed chromosome breakage was first observed by Boveri (1887) in the horse parasitic nematode *Parascaris equorum*. The single pair of germline chromosomes is broken into more than 40 pieces during development of the somatic tissue, and during this process all the heterochromatic regions of the genome are eliminated. Such chromatin diminution occurs in many species (reviewed by Pimpinelli and Goday 1989) although the molecular details have not been established. In many cases diminution is associated with chromosome fragmentation and the formation of new telomeres, at least as judged by the increase in the number of visible chromosome ends.

One of the few species which has been analysed in detail is *Ascaris lumbricoides* (Müller *et al.* 1991; reviewed by Tobler *et al.* 1992). It is not clear if the chromosome number changes in this species as the chromosomes are too small to be accurately counted. However, the heterochromatic termini of the chromosomes are removed, which suggests that new chromosome ends are produced. One example of a somatic telomere and the corresponding region of the germ-line genome have been sequenced. Breakage is accompanied by $(TTAGGC)_n$ repeat addition at the breakpoint, a sequence which does not pre-exist at this site in the germ-line genome. All the $(TTAGGC)_n$ signal is BAL31-sensitive in the somatic tissue, suggesting that chromosome breakage is associated with *de novo* telomere sequence addition. The sequence onto which the $(TTAGGC)_n$ repeats are added bears little similarity to these repeats, and as in ciliates this might indicate modifications of the substrate recognition process *in vivo*.

Despite the fact that we have almost no idea why some species undergo chromatin diminution, it provides an excellent *in vivo* experimental system to study *de novo* telomere addition. Although limited molecular details are known for a few ciliates and *Ascaris*, it seems likely that other cases of chromosome breakage may be associated with *de novo* telomere addition.

Non-programmed healing events in other species

Humans

Chromatin diminution and programmed *de novo* telomere formation do not occur in humans. However, two examples of spontaneous chromosome breakage healed by telomere addition have been reported (Wilkie *et al.* 1990; Lamb *et al.* 1993; reviewed by Broccoli and Cooke 1993). Both cases were patients with α-thalassaemia, resulting from alteration of one copy of the α-globin locus, which is located within a few hundred kilobases of the telomere of chromosome 16 (see Chapter 10). In both cases the break has been healed by *de novo* $(TTAGGG)_n$ addition; there are no $(TTAGGG)_n$ repeats pre-existing at the break site. It has been shown that an oligonucleotide corresponding to one of these breakpoint sequences acts as a primer for human telomerase *in vitro* (Morin 1991; see Chapter 4 for a more detailed discussion).

A similar example of healing by the addition of $(TTAGGG)_n$, this time in a cultured cell, has been reported by Murnane and Yu (1993). They observed a diffuse band on a Southern blot for one out of 108 human cell clones transfected with plasmid DNA which did not contain $(TTAGGG)_n$ repeats. Further analysis of this clone indicated that chromosome breakage had occurred at the site of integration of this plasmid, and that $(TTAGGG)_n$ repeats had been added onto plasmid sequences. Although the repeats are added onto non-telomeric sequences, in a way analogous to the α-thalassaemia example, it is not known if the breakpoint sequence functions as a telomerase primer *in vitro*.

Plasmodium falciparum

This malarial parasite is characterized by extensive chromosome size polymorphism between isolates from different geographical locations (see Chapter 3). Some of this polymorphism reflects what appears to be spontaneous chromosome breakage followed by *de novo* terminal repeat addition. A number of such healed spontaneous breaks have been sequenced, in particular those within a group of histidine-rich protein genes (Pologe and Ravetch 1988; Scherf and Mattei, 1992). In none are there any telomere repeats pre-existing at the breakpoints, nor is there any similarity between the sites of addition. As with *Paramecium*, the data are consistent with any double-strand break being healed, irrespective of sequence.

Saccharomyces cerevisiae

Non-telomeric ends can be healed in yeast in a number of ways. *Saccharomyces cerevisiae* is capable of very efficient homologous recombination and healing can occur by this pathway. For example, a linear vector terminating at a site within a cloned Y' element is able to heal by recombining with an endogenous telomere and acquiring a full-length Y' element, complete with its cap of TG_{1-3} repeats (Dunn *et al.* 1984). Recombination can also transfer Y' elements to chromosome termini to enable certain cells to survive the *est1Δ* mutation (Chapter 9). Both of these recombinational processes are dependent on the product of the *RAD52* gene, which is required for most types of mitotic recombination. However, there is no evidence to suggest that TG_{1-3} repeat maintenance requires such events, as normal telomere maintenance is unaffected in a *rad52* strain (Lundblad and Blackburn 1993).

A linear end terminating in a variety of TG-rich sequences can be healed by the direct addition of yeast TG_{1-3} repeats by a process which does not require *RAD52* (for example see Lustig 1992). The TG-rich sequence does not have to be at the very end; molecules with at least 30 bp of non-telomeric sequence between the TG_{1-3} repeats and the 3' end can be healed (Murray *et al.* 1988). Similar results have been found for healing *in vivo* in mammalian cells, where $(TTAGGG)_n$ repeats can be added onto non-telomeric sequences at the 3' end of a $(TTAGGG)_n$ array (Chapter 4). As described in Chapter 4, oligonucleotides composed of *Tetrahymena* TTGGGG repeats separated from the 3' terminus by over 30 bp of non-telomeric sequence are substrates for *Tetrahymena* telomerase *in vitro*, with the new $(TTGGGG)_n$ repeats being added directly onto the non-telomeric sequence. It has been suggested that telomerase can recognize internal telomeric sequences and promote sequence addition to a nearby 3' end. Despite these special cases, in general TG_{1-3} repeats are not added onto non-telomeric sequences in *S. cerevisiae*, in contrast to healing in ciliates, where any free end appears to be a substrate for telomere addition.

Sequence exchanges between the ends of a linear molecule can occur during terminal repeat addition to TG-rich sequences (Pluta and Zakian 1989; Wang and Zakian 1990; reviewed by Zakian *et al.* 1990). For example, if a molecule of the following structure:

$$5'\text{-}(A_4C_4)_n \ldots\ldots\ldots\ldots\ldots (T_2G_4)_{\sim 50}\text{-}3'$$

is introduced into yeast, the majority of the resulting transformants contain molecules which have transferred some T_2G_4 repeats from the right end to the left. These are then capped by yeast TG_{1-3} repeats, producing the following type of structure:

$$5'\text{-}(AC_{1-3})_x(A_2C_4)_{30-40}(A_4C_4)_m \ldots\ldots\ldots\ldots (T_2G_4)_{\sim 50}(TG_{1-3})_x\text{-}3'$$

This transfer is independent of the product of the *RAD52* gene, in contrast to the transfer of Y' elements between telomeres. Neither T_2G_4 nor T_4G_4 repeats have been observed to transfer to endogenous yeast telomeres during such experiments and there is no direct evidence that the final TG_{1-3} sequence addition occurs by a recombination mechanism.

TG_{1-3} repeat addition onto non-telomeric sequences is much more difficult to explain using recombinational models of telomere sequence addition such as strand invasion and gene conversion (Fig. 1.3) unless the recombination requires very little sequence homology. The strongest evidence against recombination being responsible for TG_{1-3} sequence addition comes from work by Kramer and Haber (1993). They analysed 44 examples of telomere formation at short TG-rich sequences in yeast, and found that in 37 cases one of two short motifs (GTGTGGGTGTG or GTGTGTGGGTGTG) were either the first nucleotides added or overlapped the junction between the chromosomal primer and the newly added sequences. This would be difficult to rationalize by a recombinational model where the chromosome terminus invades a telomeric TG_{1-3} array and is extended by gene conversion (Fig. 1.3) which would not be predicted to produce any similarity in the first sequence added. Healing events in yeast can probably be best explained by a telomerase activity adding GTGTGGGTGTG or GTGTGTGGGTGTG repeats. Although some sequence exchange between terminal plasmid sequences can occur during yeast terminal repeat addition onto TG-rich sequences, there is no direct evidence that this reflects a process involved either in healing or in normal telomere maintenance.

Summary

Analysis of chromosome breakage and healing in ciliates and *Ascaris* suggests a marked difference in the *in vitro* and *in vivo* substrates for telomere sequence addition. However, these examples involve a developmentally programmed breakage and healing process and may reflect the existence of special mechanisms to promote an interaction between telomerase and non-telomeric double-strand breaks in these cells (discussed in Chapter 4). In most other species non-telomeric double-strand breaks are not healed by *de novo* terminal repeat addition at a high frequency. Most cells perhaps restrict telomere sequence addition to telomere-like sequences because promiscuous healing of breaks in germ-line cells would lead to karyotypic instability.

References

Reviews

Broccoli, D. and Cooke, H. (1993). Aging, healing, and the metabolism of telomeres. *Am. J. Human Genet.*, **52**, 657–60.

Pimpinelli, S. and Goday, C. (1989). Unusual kinetochores and chromatin diminution in *Parascaris*. *Trends Genet.*, **5**, 310–15.

Prescott, D. M. (1992*a*). The unusual organization and processing of genomic DNA in hypotrichous ciliates. *Trends Genet.*, **8**, 439–45.

Prescott, D. M. (1922*b*). Cutting, splicing, reordering, and elimination of DNA sequences in hypotrichous ciliates. *Bioessays*, **14**, 317–24.

Tobler, H., Etter, A., and Müller, F. (1992). Chromatin diminution in nematode development. *Trends Genet.*, **8**, 427–32.

Zakian, V. A., Runge, K., and Wang, S.-S. (1990). How does the end begin? Formation and maintenance of telomeres in ciliates and yeast. *Trends Genet.*, **6**, 12–16.

Primary papers

Austerberry, C. F. and Yao, M.-C. (1988). Sequence structures of two developmentally regulated, alternative DNA deletion junctions in *Tetrahymena thermophila*. *Mol. Cell. Biol.*, **8**, 3947–50.

Austerberry, C. F., Snyder, R. O., and Yao, M.-C. (1989). Sequence microheterogeneity is generated at junctions of programmed DNA deletions in *Tetrahymena thermophila*. *Nucleic Acids Res.*, **17**, 7263–72.

Baroin, A., Prat, A., and Caron, F. (1987). Telomeric site position heterogeneity in macronuclear DNA of *Paramecium primaurelia*. *Nucleic Acids Res.*, **15**, 1717–28.

Boveri, T. (1887). Uber Differenzierung der Zellkerne wahrend Furchung des Eies von Ascaris megalocephala. *Anat. Anz.*, **2**, 688–93.

Dunn, B., Szauter, P., Pardue, M. L., and Szostak, J. W. (1984). Transfer of yeast telomeres to linear plasmids by recombination. *Cell*, **39**, 191–201.

Forney, J. D. and Blackburn, E. H. (1988). Developmentally controlled telomere addition in wild-type and mutant Paramecia. *Mol. Cell. Biol.*, **8**, 251–8.

Gilley, D., Preer, J. R., Aufderheide, K. J., and Polisky, B. (1988). Autonomous replication and addition of telomere-like sequences to DNA microinjected into *Paramecium tetraurelia* macronuclei. *Mol. Cell. Biol.*, **8**, 4765–72.

Herrick, G., Hunter, D., Williams, K., and Kotter, K. (1987). Alternative processing during development of a macronuclear chromosome family in *Oxytricha fallax*. *Genes Dev.*, **331**, 1047–58.

King, B. O. and Yao, M.-C. (1982). Tandemly repeated hexanucleotide at Tetrahymena rDNA free end is generated from a single copy during development. *Cell*, **31**, 177–82.

Klobutcher, L. A., Jahn, C. L., and Prescott, D. M. (1984). Internal sequences are eliminated from genes during macronuclear development in the ciliated protozoan Oxytricha nova. *Cell*, **36**, 1045-55.

Klobutcher, L. A., Vailonis-Walsh, A. M., Cahill, K., and Ribas-Aparicio, R. M. (1986). Gene-sized macronuclear DNA molecules are clustered in micronuclear chromosomes of the ciliate *Oxytricha nova. Mol. Cell. Biol.*, **6**, 3606-13.

Klobutcher, L. A., Huff, M. E., and Gonye, G. E. (1988). Alternative use of chromosome fragmentation sites in the ciliated protozoan *Oxytricha nova. Nucleic Acids Res.*, **16**, 251-64.

Kramer, K. M. and Haber, J. E. (1993). New telomeres in yeast are initiated with a highly selected subset of TG_{1-3} repeats. *Genes Dev.*, **7**, 2345-56.

Lamb, J., Harris, P. C., Wilkie, A. O. M., Wood, W. G., Dauwerse, J. G., and Higgs, D. R. (1993). De novo truncation of chromosome 16p and healing with $(TTAGGG)_n$ in the α thalassemia/mental retardation syndrome (ATR-16). *Am. J. Human. Genet.*, **52**, 668-76.

Lundblad, V. and Blackburn, E. H. (1993). An alternative pathway for yeast telomere maintenance rescues *est1⁻* senescence. *Cell*, **73**, 347-60.

Lustig, A. J. (1992). Hoogsteen G-G base pairing is dispensable for telomere healing in yeast. *Nucleic Acids Res.*, **20**, 3021-8.

Morin, G. B. (1991). Recognition of a chromosome truncation site associated with α-thalassaemia by human telomerase. *Nature*, **353**, 454-6.

Müller, F., Wicky, C., Spicher, A., and Tobler, H. (1991). New telomere formation after developmentally regulated chromosomal breakage during the process of chromatin diminution in Ascaris lumbricoides. *Cell*, **67**, 815-22.

Murnane, J. P. and Yu, L.-C. (1993). Acquisition of telomere repeat sequences by transfected DNA integrated at the site of a chromosome break. *Mol. Cell. Biol.*, **13**, 977-83.

Murray, A. W., Claus, T. E., and Szostak, J. W. (1988). Characterization of two telomeric DNA processing reactions in *Saccharomyces cerevisiae. Mol. Cell. Biol.*, **8**, 4642-50.

Pluta, A. F. and Zakian V. A. (1989). Recombination occurs during telomere formation in yeast. *Nature*, **337**, 429-33.

Pologe, L. G. and Ravetch, J. V. (1988). Large deletions result from breakage and healing of P. falciparum chromosomes. *Cell*, **55**, 869-74.

Scherf, A. and Mattei, D. (1992). Cloning and characterization of chromosome breakpoints of *Plasmodium falciparum*: breakage and new telomere formation occurs frequently and randomly in subtelomeric genes. *Nucleic Acids Res.*, **20**, 1491-6.

Spangler, E. A., Ryan, T., and Blackburn, E. H. (1988). Developmentally regulated telomere addition in *Tetrahymena thermophila. Nucleic Acids Res.*, **16**, 5569-85.

Wang, S.-S. and Zakian, V. A. (1990). Telomere-telomere recombination provides an express pathway for telomere acquisition. *Nature*, **345**, 456-8.

Wilkie, A. O. M., Lamb, J., Harris, P. C., Finney, R. D., and Higgs, D. R. (1990). A truncated human chromosome 16 associated with α thalassaemia is stabilized by addition of telomeric repeat $(TTAGGG)_n$. *Nature*, **346**, 868-71.

Yao, M.-C., Zheng, K., and Yao, C.-H. (1987). A conserved nucleotide sequence at the sites of developmentally regulated chromosomal breakage in Tetrahymena. *Cell*, **48**, 779-88.

Yao, M.-C., Yao, C.-H., and Monks, B. (1990). The controlling sequence for site-specific chromosome breakage in Tetrahymena. *Cell*, **63**, 763–72.

Yokoyama, R. and Yao, M.-C. (1986). Sequence characterization of *Tetrahymena* macronuclear DNA ends. *Nucleic Acids Res.*, **14**, 2109–22.

Yu, G.-L. and Blackburn, E. H. (1991). Developmentally programmed healing of chromosomes by telomerase in Tetrahymena. *Cell*, **67**, 823–32.

7

Human telomere loss, ageing, and cancer

The telomere is a specialized structure which causes the natural ends of a linear chromosome to behave differently from simple double-strand breaks. This concept was first elaborated by Muller (Chapter 2) on the basis of the inability to recover terminal deletions in *Drosophila*. In what way does such a double-strand break behave differently from a telomere? Such breaks in maize, resulting from the action of dicentric chromosomes, cause repeated rounds of breakage–fusion–bridge cycles (McClintock 1941, 1942). Thus the idea developed that without a telomere a chromosome end was prone to fusion and recombination. Another consequence is that a break will activate a cell cycle checkpoint and arrest cell division. In yeast the action of components of the *RAD9* pathway cause cell cycle arrest in response to DNA damage (Hartwell and Weinert 1989), so that a cell with a damaged genome does not exacerbate the problem by going through division and instead is allowed time to repair the break. A single break in the yeast genome, for example caused by the HO endonuclease, is sufficient to cause cell cycle arrest (Brown *et al.* 1991; Bennett *et al.* 1993). If a cell with linear chromosomes is to continue division then telomeres must somehow not activate such checkpoints.

There are various examples where changes in telomere structure, either quantitative or qualitative, have dramatic phenotypes. For example, telomeres in *Tetrahymena* have been manipulated using cells expressing mutant telomerase template RNAs; these cells show marked growth defects and eventually cease to divide (Yu *et al.* 1990). Both the sequence and the length of the telomeres is altered, with both cells accumulating shorter telomeres and those accumulating longer telomeres being sick, suggesting that changed telomere structure is detrimental to cell viability (see Chapter 4). Similarly, yeast cells with the *est1* mutation show progressively shorter telomeres and eventually die (Lundblad and Szostak 1989). In neither case is it clear what causes this pleiotropy and death. Although these cells could be described as senescing, there is no evidence that this is related to the senescence of human cells that is the subject of this chapter. However, it does demonstrate that altered telomere structure can have adverse effects on a cell. This chapter addresses the question of what might happen if a human cell were to lose one or more telomeres, and the potential relationship between telomere loss, ageing, and cancer.

Telomere loss in cultured human cells

If a cell does not possess a mechanism to add DNA to its telomeres to over-come the end-replication problem, then it will progressively lose DNA from the ends of its chromosomes with successive divisions. A detailed mathe-matical analysis of the relationship between the end-replication problem and the kinetics of telomere loss can be found in Levy *et al.* (1992). The possibility that human somatic cells might show telomere loss stemmed from the observation that human telomeres are longer in sperm than in blood (Cooke and Smith 1986; Allshire *et al.* 1988, 1989; Cross *et al.* 1989; de Lange *et al.* 1990; Hastie *et al.* 1990). Progressive loss of telomeric DNA in somatic tissue has indeed been found (reviewed by Greider 1990; Broccoli and Cooke 1993). This may reflect telomerase expression being limited to germ-line tissues, although as yet there is little direct evidence for this.

What might happen to a human cell if it loses a telomere? The possibility that telomere loss might lead to cell senescence was first put forward by Olovnikov (1971, 1973) where telomere loss was suggested to cause a lethal deletion of some telomere-adjacent gene. The current model suggests that telomere loss reveals a non-telomere free DNA end which in turn signals cell cycle arrest. Another possibility is that telomere loss causes genome instability, which in turn might lead to cancer.

Human telomere length is usually measured by Southern analysis. Genomic DNA is digested with a restriction enzyme which recognizes a 4 bp sequence, and the length of fragments which hybridize to a probe that detects the human terminal repeat sequence, $(TTAGGG)_n$, is measured. By using enzymes which do not cleave within the $(TTAGGG)_n$ arrays but do cleave frequently elsewhere in the human genome, the size of these frag-ments provides a reasonable estimate of telomere length. Densitometric quantification of the $(TTAGGG)_n$ signal, which is a smear owing to length heterogeneity, can also be used.

Human fibroblasts undergo a finite number of divisions in culture and then senesce (reviewed by Goldstein 1990). The senescence of these cells may have a role in ageing of the organism, especially in those areas of the body which have a high degree of cellular turnover. Harley *et al.* (1990) have passaged a primary culture of normal, untransformed human fibro-blasts *in vitro*, taking samples at various stages until the cultures senesced and ceased division. As the cultures underwent successive rounds of division *in vitro*, telomere length decreased at a rate of 48 ± 21 bp per population doubling (Harley *et al.* 1990; Levy *et al.* 1992). It was speculated that telomere shortening might have a causal role in this senescence.

To date, the available data are consistent with arrays of $(TTAGGG)_n$ being sufficient to stabilize a human chromosome by providing essential telomere function. Thus loss of $(TTAGGG)_n$ might reasonably be inter-preted as loss of telomere function. One expectation is that loss of a telomere would lead to an increased rate of recombination with other chromosomes, leading to formation of dicentric chromosomes and classical breakage–fusion–bridge cycles, and in turn genome rearrangements and aneuploidy. Such genome scrambling could lead to loss of genes and a cessation of cell division. Telomere loss also produces what is in effect a double-strand break. As in yeast, checkpoints in the mammalian cell cycle may cause such a double-strand break in a chromosome to signal cell cycle arrest (Hartwell and Weinert 1989; Brown *et al.* 1991; Roberge 1992; Murray 1993).

Human fibroblast senescence *in vitro*

After a certain number of cell divisions in culture, primary human cells undergo senescence and cease to divide. This point has been termed M_1 (see Shay *et al.* 1991, for a review on senescence and transformation). Fibroblasts transformed with a DNA tumour virus such as simian virus 40 (SV40) acquire an extended life-span amounting to an additional 20–30 population doublings *in vitro* until a second point, termed M_2, is reached. At this point cell numbers remain the same as division is matched by an equal rate of cell death ('crisis'). After crisis cell numbers decrease but rare ($< 10^{-6}$) foci of growing cells can develop, and these give rise to immortal cell lines.

One hypothesis is that M_1 is a cell cycle checkpoint that SV40 T antigen overcomes and this can be reasserted if T antigen is removed. This has been shown by experiments using either a SV40 T antigen gene driven by an inducible promoter, or one encoding a temperature-sensitive protein (Radna *et al.* 1989; Wright *et al.* 1989). If the protein is removed cells will either continue to divide until they reach the M_1 point or stop dividing immediately if they are already between M_1 and M_2 (Ide *et al.* 1984; Radna *et al.* 1989; Wright *et al.* 1989). Particularly fascinating is that even though M_1 does not occur in the presence of T antigen the 'clock' is still running, so that when T antigen is removed the cell has been counting cell divisions and reaches M_1 at the same point. This implies that M_1 is a mechanism that is reacting to some progressive cellular clock that is bypassed but not inactivated by T antigen. It is unclear how cells measure time and count divisions but there are many speculations (Groves *et al.* 1991).

Cells may cease division for many different reasons. Damage to the cell could occur in a variety of ways, such as protein degradation, free radical damage, accumulation of waste metabolic products, and so forth (reviewed by Dice 1993). Cytological abnormalities such as dicentric chromosomes, telomere associations, and aneuploidy increase dramatically in the final, senescent stages of the growth of primary cultures *in vitro* (Harley 1991). One view is that division ceases because the cell has become so damaged that it is no longer physically capable of DNA replication and mitosis.

However, senescent cells can re-enter the cell cycle if transformed with SV40 T antigen, indicating that the ability to grow and divide remains, but is suppressed. That a single gene is able to reactivate cell division suggests that an active signal to cease division is being produced by senescent cells which is bypassed in the presence of SV40 T antigen. This is supported by the dominance of the senescent phenotype when fusing such cells with immortal transformed cells; a senescent fibroblast will inhibit the initiation of DNA synthesis in a young nucleus in a heterokaryon, suggesting that the cessation of division is the result of a dominant diffusible signal. An attractive hypothesis is that although insufficient genetic damage is occurring to affect adversely the function of proteins being produced, continued DNA damage eventually produces so many lesions that a cell cycle checkpoint is activated and cell division is arrested. Examples of the type of damage which might accumulate are the pyrimidine base derivatives 3-methylthymine and 3-methylcytosine, which are produced by nonenzymatic methylation by *S*-adenosylmethionine. There appears to be no active repair of N^3-methylated pyrimidines by DNA glycosylases in mammalian cells, and unless these derivatives are removed by the nucleotide-excision pathway they could accumulate over time and perhaps contribute to cellular senescence (Lindahl 1993).

A control mechanism may have evolved to prevent cells with damaged chromosomes continuing to divide. This in turn could inhibit the genome alterations likely to be a prelude to tumorigenesis. The product of the p53 tumour-suppressor gene, which is deleted or mutated in at least half of all human cancers, may be involved in this control (Prives 1993). p53 function is inhibited by the products of various DNA tumour virus gene products, such as SV40 T antigen (Perry and Levine 1993), and this may be how SV40 T antigen overcomes M_1 arrest.

p53 appears to be involved in the maintenance of a stable karyotype during somatic cell growth. It is involved in the suppression, or selection against, a range of genome rearrangements which might lead to changes in the karyotype; one of the easiest to measure is gene amplification. For example, normal diploid fibroblasts have a stable karyotype and do not undergo gene amplification in culture at any detectable frequency ($<10^{-9}$). In contrast, transformed cells readily undergo gene amplification in response to drug selection (10^{-3}–10^{-5}). The ability to amplify genes is a recessive trait in cell fusion experiments, implying normal cells have a factor that suppresses it. Loss of p53 is sufficient to permit a cell to undergo gene amplification in culture (Livingstone *et al.* 1992).

One clear example is the behaviour of cells in culture from mice with a homozygous null mutation of the p53 gene. Wild-type cells do not undergo gene amplification and retain a predominantly diploid karyotype. In contrast, cells homozygous for the p53 null mutation are permissive for gene amplification and also become progressively more aneuploid (Livingstone *et al.* 1992). Mice homozygous for such a null mutation are viable but tend to develop cancer at an early age (Donehower *et al.* 1992). It seems likely that p53 acts as a tumour suppressor by being part of a cell cycle checkpoint that inhibits cell division in response to genome damage. p53 may function by monitoring the state of DNA before entry into S phase, as fibroblasts lacking p53 do not arrest in G_1 in response to γ-irradiation. Thus in response to genome damage cells that have sustained mutations are prevented from further proliferation or even triggered to undergo cell death by apoptosis (Lane 1993; Lu and Lane 1993; Perry and Levine 1993).

Human telomeres shorten with age *in vivo*

The senescence of human fibroblasts may have a role in ageing *in vivo* for humans, although its exact relationship to organismal ageing is still far from clear (Goldstein 1990; Groves *et al.* 1991; McCormick and Campisi 1991). Telomere loss with division *in vitro* for human fibroblasts predicts that a similar loss might be seen for cells *in vivo*. It is therefore of interest to consider those cell types in the human body which undergo many rounds of division during life. Hastie *et al.* (1990) have studied human lymphocytes and find a significant reduction in telomere length with increasing donor age, the rate of loss being about 33 bp per year. Similarly, Lindsey *et al.* (1991) and Allsopp *et al.* (1992) find telomere sequence loss (approximately 20 bp per year) with age in skin samples. It is noteworthy that although the results are statistically significant, there is a large scatter on plots of telomere length with age and the correlation is rather weak. This may be because donor age is not an absolute predictor of the number of divisions a cell has undergone. This is supported by data from Allsopp *et al.* (1992) who have established primary fibroblast cultures from donors of varying age. They find that there is indeed a large amount of scatter on plots of donor age versus the life-span of the culture *in vitro*. However, there is a very strong correlation between remaining life-span and initial telomere length, with shorter initial telomeres found in those cultures with shorter life-spans.

The chromosome abnormalities seen in senescent cells *in vitro* would be predicted to be found in equivalent cells *in vivo*. Indeed, dicentrics are the only chromosomal aberration to increase significantly with age *in vivo* (Bender *et al.* 1989).

The conclusions from these studies are that telomeres do shorten with age *in vivo*, and that telomere length is an excellent biomarker for the potential number of cellular divisions remaining. Telomere length can be viewed, therefore, as recording the history of the cell: the shorter the telomeres, the more divisions the cell is likely to have undergone. Thus telomere length can be a useful biomarker for studying the ageing processes *in vivo*.

Certain human syndromes lead to premature ageing, and the replicative capacity of fibroblasts in culture from such individuals is markedly reduced. Telomeres of patients with one such syndrome, Hutchinson-Gilford progeria, are significantly shorter than those of age-matched controls, and the reduction in telomere length correlates with the reduced proliferative capacity of their fibroblasts *in vitro* (Allsopp *et al.* 1992). The rate of telomere loss per cell division is not significantly different from that in normal fibroblasts, however, which argues that the syndrome does not result from an abnormally high rate of telomere sequence loss. Instead, it may be that telomere loss is reflecting a higher rate of cell turnover in these patients, demonstrating how telomere length can be a useful biomarker for ageing. Similarly, Down syndrome is accompanied by premature senescence of the immune system, and Vaziri *et al.* (1993) have found that in lymphocytes, telomeres are significantly shorter in such patients for their age. In this case the rate of telomere loss per cell division *in vitro* has not been measured, and it remains unknown if the shorter telomeres reflect a higher rate of sequence loss or simply a greater turnover rate for immune cells in these patients.

Telomerase and cell immortality

That telomeres get shorter with successive rounds of division because of a lack of telomerase is strongly supported by evidence from Counter *et al.* (1992). Human embryonic kidney (HEK) cells grown *in vitro* only undergo about 20 divisions before cessation of growth. However, after transformation by SV40 T antigen they acquire an extended life-span of over 60 divisions. At this point they go into crisis and most cells do not survive this period. Those rare cells that do are now phenotypically immortal.

As HEK cells divide, their telomeres shorten and this continues to happen after transformation with SV40 T antigen. Telomere loss continues to occur even after acquiring an extended life-span, and crisis occurs at a point when almost all the telomeric DNA has been lost. Just before crisis there is a sharp rise in the frequency of chromosomal aberrations such as dicentrics. The immortal cell lines that survive crisis no longer show telomere sequence loss. Telomerase activity in cells at various stages was measured using an *in vitro* assay, and indicates that the enzyme is not active in the untransformed or extended life-span cells, but is active in the immortal cells (Counter *et al.* 1992). A similar post-crisis stabilization of telomere length has been observed in human cells transformed with HPV, although it is not known if this correlates with activation of telomerase (Klingelhutz *et al.* 1994).

In summary, the data suggest that although transformation by SV40 T antigen leads to an extended life-span it does not activate telomerase, and telomeres continue to shorten. When almost all the $(TTAGGG)_n$ signal has been lost, a marked increase in cytogenetic abnormalities occurs and the cells undergo crisis. The rare immortal cell lines that survive this period now express telomerase and no longer show shortening telomeres. This is consistent with M_2 mortality, the crisis and cell death after transformation, resulting from telomere loss.

Shortened telomeres are found in some human tumours

A number of studies have compared telomere length in human tumours with that in non-tumour tissue from the same organ. In some cases significantly shorter telomeres have been found, including colorectal carcinoma (Hastie *et al.* 1990), Wilm's tumour (de Lange *et al.* 1990), renal cell carcinoma (Bock *et al.* 1993; Holzmann *et al.* 1993; Mehle *et al.* 1994) and childhood leukaemia (Adamson *et al.* 1992). If a large tumour has arisen from a single cell by clonal expansion then the cells of that tumour will have undergone significantly more divisions than the equivalent tissue from which it came. For example, to make even a physically tiny tumour of 10^6 cells requires about 20 divisions. So although there could be an increased rate of telomere loss in tumour cells, their shortened telomeres may simply be the consequence of the extra divisions these cells have undergone relative to their parental cells. However, not all tumours have significantly shorter telomeres than their parental tissue (Hastie *et al.* 1990; Bock *et al.* 1993).

Tumour cells are often characterized by cytological abnormalities, including gene amplification, aneuploidy, and an increased frequency of dicentric chromosomes. In particular, some cancers show frequent visible attachments between the ends of chromosomes, termed telomere associations (Pathak *et al.* 1988). Tumour cells are not unique in showing telomere associations, which can also be found in patients with benign conditions such as Thiberge–Weissenbach syndrome (Dutrillaux *et al.* 1978). Where it has been investigated, cancer cells with telomere associations also show reduced telomere length (Holzmann *et al.* 1993). One possible interpretation is that telomere loss leads to the formation of highly recombinogenic free chromosome ends in these cells (reviewed by Hastie and Allshire 1989). This simple view is complicated by the observations of Saltman *et al.* (1993) who report two tumour cell lines which show severely shortened telomeres and a high frequency of telomere associations, but who also report two cell lines with similarly short telomeres which do not show telomere associations.

Can cancer be caused by telomere loss?

As a cell gets towards the end of its replicative life-span, telomere loss may result in genome rearrangements, which might be viewed as a sudden burst of mutagenesis. Such alterations could in principle facilitate the formation of a transformed cell. One particular mechanism whereby the consequences of telomere loss might be involved in cellular transformation is chromosome loss, as loss of heterozygosity (reflecting the loss of one chromosome homologue) is seen in many types of cancer. A recessive loss-of-function mutation in one allele of a tumour-suppressor locus revealed by chromo-some loss can result in transformation (Sager 1989). This mechanism would seem of particular relevance for cancers which arise in those rapidly dividing and regenerating tissues, such as skin or intestinal mucosa, predicted to show progressive telomere loss with age. However, telomere loss in many cancers, especially those of tissues such as the central nervous system which show little cellular turnover, may instead reflect the extra divisions undergone by the tumour cells (for further discussion, see above and Jankovic *et al.* 1991).

Counter *et al.* (1992) speculate that the finite life-span of SV40-transformed cells lacking telomerase might explain the frequent regression of tumours after limited growth *in vivo*. Many benign tumours naturally regress; they may lack telomerase and undergo the equivalent of *in vitro* crisis when their telomeres become critically shortened. In contrast, the cells of malignant, perhaps metastatic, tumours might express telomerase and be immortal. It will be of great interest to measure telomerase activity in a range of tumours and normal tissues; if this speculation is correct, anti-telomerase drugs might be effective against such tumours. Indeed, although telomerase activity has not been detected in cultured primary human cells, it has been detected in metastatic ovarian carcinoma cells (Counter *et al.* 1994). It should be noted, however, that although telomere loss and the M_2 mechanism may be relevant to tumour regression, the M_1 mechanism is the cause of senescence of primary cells *in vitro* and presumably *in vivo*.

Telomeres and cellular senescence

It is not clear how the senescent phenotype of cells in culture correlates with telomere loss. It has been postulated that telomere loss can explain both M_1 and M_2 mortality phases (Harley 1991; Denis and Lacroix 1993). One possibility is that M_1 reflects the activation of a cell cycle checkpoint by the loss of one or more telomeres. Telomere loss could also be linked to cell cycle arrest by deleting genes located subtelomerically. Certainly the telomere shortening in EBV-transformed human B lymphocytes can extend into subtelomeric sequences (Guerrini *et al.* 1993). Alternatively, Wright and Shay (1992) have speculated that telomere shortening could cause hypo-thetical telomere-associated repressive chromatin to spread further along the chromosome and inactivate subtelomeric genes. This suggestion is based on the discovery of position effect variegation at yeast telomeres (see Chapter 8). However, genes can be expressed when close to newly created human telomeres (Farr *et al.* 1991, 1992; Itzhaki *et al.* 1992; Barnett *et al.* 1993; Bayne *et al.* 1994). In yeast long telomeres actually impose increased repression of adjacent genes (Kyrion *et al.* 1993) and thus shortened telomeres might lead to the activation of a subtelomeric gene which in turn could produce the dominant cell cycle arrest signal. Nevertheless, the ability of double-strand breaks to activate cell cycle checkpoints suggests that non-telomeric free ends caused by telomere loss may in themselves cause cell cycle arrest. For M_2 arrest it may be reasonable to assume that pleiotropy associated with the genome rearrangements which occur is sufficient to cause a cessation of division.

Cell mortality and the germ:soma division

The cells of a modern-day animal are descended by division from the very first eukaryotic cells. Cells of the germ-line appear to have maintained the reproductive immortality their unicellular ancestors possessed. In contrast, somatic cells are able to undergo only a finite number of cell divisions and seem to have acquired mortality. Senescent cells are dominant over most types of transformed cells in fusion experiments, the exception being those transformed with DNA tumour viruses such as SV40 (Denis and Lacroix 1993; Dice 1993), indicating that mortality in somatic cells is a gain of function.

In the early stages of metazoan evolution a limited cellular life-span may have evolved as a crude but effective way of limiting the size of the organism, although this is not the main determinant of body size in modern metazoans. The restricted division potential of somatic cells, an extra layer of control imposed on otherwise immortal cells, may now be used to provide a degree of protection against life-threatening unlimited cellular proliferation. It seems likely that the increase in life-span, if any, that might be conferred by immortal somatic cells would be abrogated by an increased risk of cancer. Furthermore, limited growth potential does not necessarily have a detrimental effect on an organism if it is sufficient to permit the animal to reach its optimum life-span, which can be largely determined by extrinsic factors (see below).

Telomerase activity has been detected in transformed human cells grown *in vitro* and in metastatic ovarian carcinoma cells (Morin 1989; Counter *et al.* 1992, 1994). As yet few analyses have been performed on human tissue, as opposed to cultured cells, and there is no direct evidence that telomerase is active in human germ-line tissue. However, unlike the various somatic cells described above, human sperm samples do not show a decrease in telomere length with donor age. This is consistent with the hypothesis that telomerase and telomere length maintenance are active in germ-line cells. Furthermore, in maize broken chromosomes are healed only in sporophytic (germ-line) tissue and not in the endosperm, which is differentiated somatic tissue (McClintock 1941, 1942). This again suggests that telomerase is active in germ-line cells. The germ-line transmission of a spontaneous healed chromosome 16 break in humans (see Chapter 6) also argues that telomere addition, and thus probably telomerase, is active in human germ-line tissues.

The biology of organismal ageing

The causes of death for animals in the wild are very diverse (Finch 1990). Many herbivores such as horses die of starvation after their teeth have worn down, whereas cancer and cardiovascular disease are major causes of human death in developed countries. Not all species appear to age, in that they do not all show an age-related increase in mortality rates; the lobster is one example (Finch 1990). However, if an organism does age it is likely to become more prone to death by accident, disease, or being killed by a predator. For example, decreased mobility might reduce the ability to flee from predators and deterioration of the immune system would reduce resistance to pathogens. Therefore, although the final causes of death may vary, ageing and its common cellular changes may underlie many of these causes.

Early theories regarding the evolution of ageing postulated some form of group selection (Kirkwood 1989). They postulate that ageing, and therefore death, is being actively selected *for*, perhaps so that parents do not continue to compete with their offspring for limited food supplies. However, senescence is rare in the wild, with most animals dying of other causes before reaching old age. Although programmed death for the benefit of the species may occur in some cases, one must question whether ageing serves any useful purpose for most species. Many modern theories argue instead that ageing is an unavoidable by-product of optimizing other biological processes. Ageing is not of benefit *per se*, and it is longevity rather than ageing that has been selected for. Ageing can be the result of optimizing early reproductive success, reflecting the fact that life is one long series of compromises (Dice 1993; Partridge and Barton 1993).

Any theory of ageing must be able to account for observational data (Kirkwood 1989; Finch 1990; Partridge and Barton 1993). For example, birds generally have greater maximum life-spans than mammals. Flight may provide some degree of protection against predators, and so explain the longer life-span of birds. This is supported by the facts that bats are particularly long-lived mammals and flightless birds such as ostriches are unusually short-lived among birds. Similarly, turtles and tortoises outlive other reptiles, and their shells may provide added protection against predation. This difference in life-spans holds even in the absence of predators, such as when in captivity. Intrinsic life-span may therefore have been adapted to the rates of predation experienced in the wild.

The thrust of most animals' lives is to find as much food as possible, avoid being eaten themselves or dying by accident, and produce as many offspring as possible which reach sexual maturity. Resources, in the form of energy taken in as food, can be used for a number of purposes. It can be used to give rapid early growth, so that the animal is soon able to fend for itself. It can also be used to produce offspring and help them achieve sexual maturity, either directly by contributing to the tissue of the embryo or by post-natal care and feeding.

Resources can also be used to repair the fabric of the animal's body. The tissue of an animal's body does not last indefinitely. There are mechanisms to repair or renew many macromolecules, and these require energy. In particular, environmental mutagens, whether they be chemicals or radiation, are constantly attacking chromosomal DNA, and genetic damage can seriously affect the life-span of an organism if it leads to life-threatening cancer. Indeed, even without extrinsic factors, DNA itself will degrade because of the inherent thermodynamic properties of covalent bonds. These processes are combated by various energy-dependent pathways of DNA repair (Lindahl 1993).

How resources are allocated depends on the life-style to which the animal has adapted, and this can ultimately determine the life-span of the organism. For example, an organism that is subject to high rates of predation might devote little of its resources to repairing its body (a 'disposable soma' theory). After all, there is little benefit in having a body that will last for a hundred years if you are eaten within the year. The energy saved can be applied to producing a greater number of offspring in the short time that is usually available.

The repair systems for molecules such as DNA and for the general structure of the cell may therefore be imperfect. The energy saved can be devoted to more rapid early growth and greater reproduction. The ageing that will ensue in this case may not be very strongly selected against, if the animal usually dies of disease or at the hands of a predator. Resources need to be expended on somatic repair only so as to ensure that ageing does not occur so rapidly that it becomes a significant cause of death relative to other mechanisms.

The compromises made will depend on the species. For a rodent the main aim might be to maximize reproduction in the face of a high rate of predation. In contrast, a species not prone to predation may attempt to maximize its life-span by devoting more resources to somatic repair. Here the main aim might be to maintain efficient DNA repair and a low level of tumorigenesis. Indeed, senescence itself may have evolved because the limited replicative potential of somatic cells suppresses tumorigenesis. Telomere loss and M_2 mortality may have evolved as a second line of protection against tumorigenesis in long-lived species whose life-span is more affected by cancer.

Summary

The current model can be summarized as follows. Lack of telomerase activity in primary human cells leads to telomere shortening. Eventually the loss of all the $(TTAGGG)_n$ at one telomere reveals an end which is no longer recognized as a telomere by the cell. Instead it behaves as a double-strand break and activates a cell cycle checkpoint, seen as M_1 mortality. As this can be overcome by expression of SV40 T antigen, this suggests that T antigen bypasses the dominant signal that triggers cell cycle arrest at this point. However, telomerase remains inactive, and telomeres continue to shorten. As more and more telomeres become lost the incidence of chromosomal rearrangements resulting from the fusigenic tendency of double-strand breaks increases, leading to deleterious effects on the ability to grow and divide. The effect of this is a crisis and M_2 mortality. To enable continued cell division, telomere loss must cease, requiring the activation of telomerase. Therefore only those cells which have activated telomerase can be immortal. Nevertheless, despite being a plausible model for the senescence of human cells *in vitro*, the evidence for the involvement of telomeres in cellular senescence and transformation is largely circumstantial and not proven at present.

One role of M_2 mortality may be as a mechanism to protect against cancer, at least through our reproductive period. Cancer is a disease primarily of late life in humans, and given that the average life-span in human populations has until relatively recently been quite short, cancer probably had little effect on mortality rates prior to modern times. The rate of telomere loss in humans is such that formation of a large tumour may require the reactivation of telomerase, because without it the deleterious effects of large-scale telomere loss would prevent continued growth of the tumour. Therefore, by requiring both a bypass of M_1 and reactivation of telomerase, life-threatening tumours may be harder to create, and thus take longer to arise by chance mutation.

The evidence that telomeres are involved in the M_1 mechanism which causes non-transformed cells to cease division is less convincing. It is this senescence that is presumably relevant to the biological process of cellular senescence *in vivo*. Even if true, ageing of cells, tissues, and organs occurs for many different reasons, and it is uncertain to what extent even M_1, the more biologically relevant of the two mechanisms, contributes to organismal ageing.

One implication of this hypothesis is that senescence and cellular transformation may be inextricably linked. One intriguing question is why telomerase appears to be inactive in somatic cells. If telomere loss does indeed provide an ultimate restriction to life-span, surely an animal whose somatic cells express telomerase would live longer? The answer to this may be that it would remove the M_2 mechanism and thus the formation of tumours would be greatly facilitated, so much so that it might negate any increase in life-span resulting from reduced cellular ageing. Life is just one long compromise.

References

Reviews

Broccoli, D. and Cooke, H. (1993). Aging, healing, and the metabolism of telomeres. *Am. J. Human Genet.*, **52**, 657–60.

Denis, H. and Lacroix, J.-C. (1993). The dichotomy between germ line and somatic line, and the origin of cell mortality. *Trends Genet*, **9**, 7–11.

Dice, J. F. (1993). Cellular and molecular mechanisms of aging. *Physiol. Rev.*, **73**, 149–59.

Finch, C. E. (1990). *Longevity, senescence, and the genome*. University of Chicago Press, Chicago.

Goldstein, S. (1990). Replicative senescence: the human fibroblast comes of age. *Science*, **249**, 1129–33.

Greider, C. W. (1990). Telomeres, telomerase and senescence. *Bioessays*, **12**, 363–9.

Groves, A. K., Bögler, O., Jat, P. S., and Noble, M. (1991). The cellular measurement of time. *Curr. Opin. Cell Biol.*, **3**, 224–9.

Harley, C. B. (1991). Telomere loss: mitotic clock or genetic time bomb? *Mutat. Res.*, **256**, 271–82.

Hartwell, L. H. and Weinert, T. A. (1989). Checkpoints: controls that ensure the order of cell cycle events. *Science*, **246**, 629–34.

Hastie, N. D. and Allshire, R. C. (1989). Human telomeres: fusion and interstitial sites. *Trends Genet.*, **5**, 326–30.

Kirkwood, J. L. (1989). Evolution and ageing. *Genome*, **31**, 398–405.

Lane, D. P. (1993). A death in the life of p53. *Nature*, **362**, 786–7.

Levy, M. Z., Allsopp, R. C., Futcher, A. B., Greider, C. W., and Harley, C. B. (1992). Telomere end-replication problem and cell aging. *J. Mol. Biol.*, **225**, 951–60.

Lindahl, T. (1993). Instability and decay of the primary structure of DNA. *Nature*, **362**, 709–15.

McCormick, A. and Campisi, J. (1991). Cellular aging and senescence. *Curr. Opin. Cell Biol.*, **3**, 230–34.

Murray, A. W. (1993). Sunburnt fission yeast. *Nature*, **363**, 302.

Partridge, L. and Barton, N. H. (1993). Optimality, mutation and the evolution of ageing. *Nature*, **362**, 305–11.

Perry, M. E. and Levine, A. J. (1993). Tumor-suppressor p53 and the cell cycle. *Curr. Opin. Genet. Dev.*, **3**, 50–54.

Prives, C. (1993). Doing the right thing: feedback control and p53. *Curr. Opin. Cell Biol.*, **5**, 214–18.

Roberge, M. (1992). Checkpoint controls that couple mitosis to completion of DNA replication. *Trends Cell Biol.*, **2**, 277–81.

Sager, R. (1989). Tumor suppressor genes: the puzzle and the promise. *Science*, **246**, 1406–12.

Shay, J. W., Wright, W. E., and Werbin, H. (1991). Defining the molecular mechanisms of human cell immortalization. *Biochim. Biophys. Acta*, **1072**, 1–7.

Wright, W. E. and Shay, J. W. (1992). Telomere positional effects and the regulation of cellular senescence. *Trends Genet.*, **8**, 193–7.

Primary papers

Adamson, D. J. A., King, D. J., and Haites, N. E. (1992). Significant telomere shortening in childhood leukemia. *Cancer Genet. Cytogenet.*, **61**, 204–6.

Allshire, R. C., Gosden, J. R., Cross, S. H., Cranston, G., Rout, D., Sugawara, N., *et al.* (1988). Telomeric repeat from *T. thermophila* cross hybridizes with human telomeres. *Nature*, **332**, 656–9.

Allshire, R. C., Dempster, M., and Hastie, N. D. (1989). Human telomeres contain at least three types of G-rich repeat distributed non-randomly. *Nucleic Acids Res.*, **17**, 4611–27.

Allsopp, R. C., Vaziri, H., Patterson, C., Goldstein, S., Younglai, E. V., Futcher, A. B., *et al.* (1992). Telomere length predicts replicative capacity of human fibroblasts. *Proc. Natl. Acad. Sci. USA*, **89**, 10114–18.

Barnett, M. A., Buckle, V. J., Evans, E. P., Porter, A. C. G., Rout D., Smith, A. G., and Brown, W. R. A. (1993). Telomere directed fragmentation of mammalian chromosomes. *Nucleic Acids Res.*, **21**, 27–36.

Bayne, R. A. L., Broccoli, D., Taggart, M. H., Thomson, E. J., Farr, C. J., and Cooke, H. J. (1994). Sandwiching of a gene within 12 kb of a functional telomere and alpha satellite does not result in silencing. *Human Mol. Genet.*, **3**, 539–46.

Bender, M. A., Preston, R. J., Leonard, R. C., Pyatt, B. E., and Gooch, P. C. (1989). Chromosomal aberration and sister-chromatid exchange frequencies in peripheral blood lymphocytes of a large human population sample. *Mutat. Res.*, **212**, 149–54.

Bennett, C. B., Lewis, A. L., Baldwin, K. K., and Resnick, M. A. (1993). Lethality induced by a single site-specific double-strand break in a dispensable yeast plasmid. *Proc. Natl. Acad. Sci. USA*, **90**, 5613–17.

Bock, S., Epplen, J. T., Noll-Puchta, H., Rotter, M., Höfler, H., Block, T., *et al.* (1993). Detection of somatic changes in human renal cell carcinomas with oligonucleotide probes specific for simple repeat motifs. *Genes Chromosomes Cancer*, **6**, 113–17.

Brown, M., Garvik, B., Hartwell, L., Kadyk, L., Seeley, T., and Weinert, T. (1991). Fidelity of mitotic chromosome transmission. *Cold Spring Harbor Symp. Quant. Biol.*, **56**, 359–65.

Cooke, H. J. and Smith, B. A. (1986). Variability at the telomeres of the human X/Y pseudoautosomal region. *Cold Spring Harbor Symp. Quant. Biol.*, **51**, 213–19.

Counter, C. M., Avilion, A. A., LeFeuvre, C. E., Stewart, N. G., Greider, C. W., Harley, C. B., and Bacchetti, S. (1992). Telomere shortening associated with chromosome instability is arrested in immortal cells which express telomerase activity. *EMBO J.*, **11**, 1921–9.

Counter, C. M., Hirte, H. W., Bacchetti, S., and Harley, C. B. (1994). Telomerase activity in human ovarian carcinoma. *Proc. Natl. Acad. Sci. USA*, **91**, 2900–4.

Cross, S. H., Allshire, R. C., McKay, S. J., McGill, N. I., and Cooke, H. J. (1989). Cloning of human telomeres by complementation in yeast. *Nature*, **338**, 771–4.

de Lange, T., Shiue, L., Myers, R. M., Cox, D. R., Naylor, S. L., Killery, A. M., and Varmus, H. E. (1990). Structure and variability of human chromosome ends. *Mol. Cell. Biol.*, **10**, 518–27.

Donehower, L. A., Harvey, M., Slagle, B. L., McArthur, M. J., Montgomery, C. A., Butel, J. S., and Bradley, A. (1992). Mice deficient for p53 are developmentally normal but susceptible to spontaneous tumours. *Nature*, **356**, 215–21.

Dutrillaux, B., Croquette, M. F., Viegas-Pequignot, E., Aurias, A., Coget, J., Couturier, J., and Lejeune, J. (1978). Human somatic chromosome chains and rings. *Cytogenet. Cell Genet.*, **20**, 70–77.

Farr, C., Fantes, J., Goodfellow, P., and Cooke, H. (1991). Functional reintroduction of human telomeres into mammalian cells. *Proc. Natl. Acad. Sci. USA*, **88**, 7006–10.

Farr, C. J., Stevanovic, M., Thomson, E. J., Goodfellow, P. N., and Cooke, H. J. (1992). Telomere-associated chromosome fragmentation: applications in genome manipulation and analysis. *Nature Genet.*, **2**, 275–82.

Guerrini, A. M., Camponeschi, B., Ascenzioni, F., Piccolella, E., and Donini, P. (1993). Subtelomeric as well as telomeric sequences are lost from chromosome in proliferating B lymphocytes. *Human Mol. Genet.*, **2**, 455–60.

Harley, C. B., Futcher, A. B. and Greider, C. W. (1990). Telomeres shorten during ageing of human fibroblasts. *Nature*, **345**, 458–60.

Hastie, N. D., Dempster, M., Dunlop, M. G., Thompson, A. M., Green, D. K., and Allshire, R. C. (1990). Telomere reduction in human colorectal carcinoma and with ageing. *Nature*, **346**, 866–8.

Holzmann, K., Blin, N., Welter, C., Zang, K. D., Seitz, G., and Henn, W. (1993). Telomeric associations and loss of telomeric DNA repeats in renal tumors. *Genes Chromosomes Cancer*, **6**, 178–81.

Ide, T., Tsuji, Y., Nakashima, T., and Ishibashi, S. (1984). Progress of aging in human diploid cells transformed with a tsA mutant of simian virus 40. *Exp. Cell Res.*, **150**, 321–8.

Itzhaki, J. E., Barnett, M. A., MacCarthy, A. B., Buckle, V. J., Brown, W. R. A., and Porter, A. C. G. (1992). Targeted breakage of a human chromosome mediated by cloned human telomeric DNA. *Nature Genet.*, **2**, 283–7.

Jankovic, G. M. J., Colovic, M. D., and Petrovic, M. D. (1991). Telomere loss and cancer. *Nature*, **350**, 197.

Klingelhutz, A. J., Barber, S. A., Smith, P. P., Dyer, K., and McDougall, J. K. (1994). Restoration of telomeres in human papillomavirus-immortalized human anogenital epithelial cells. *Mol. Cell. Biol.*, **14**, 961–9.

Kyrion, G., Liu, K., Liu, C., and Lustig, A. J. (1993). RAP1 and telomere structure regulate telomere position effects in *Saccharomyces cerevisiae*. *Genes Dev.*, **7**, 1146–59.

Lindsey, J., McGill, N. I., Lindsey, L. A., Green, D. K. and Cooke, H.J. (1991). In vivo loss of telomeric repeats with age in humans. *Mutat. Res.*, **256**, 45–8.

Livingstone, L. R., White, A., Sprouse, J., Livanos, E., Jacks, T., and Tisty, T. D. (1992). Altered cell cycle arrest and gene amplification potential accompany loss of wild-type p53. *Cell*, **70**, 923–35.

Lu, X. and Lane, D. P. (1993). Differential induction of transcriptionally active p53 following UV or ionizing radiation: defects in chromosome instability syndromes? *Cell*, **75**, 765–78.

Lundblad, V. and Szostak, J. W. (1989). A mutant with a defect in telomere elongation leads to senescence in yeast. *Cell*, **57**, 633–43.

McClintock, B. (1941). The stability of broken ends of chromosome in *Zea mays*. *Genetics*, **26**, 234–82.

McClintock, B. (1942). The fusion of broken ends of chromosomes following nuclear fusion. *Proc. Natl. Acad. Sci. USA*, **28**, 458–63.

Mehle, C., Ljungberg, B., and Roos, G. (1994). Telomere shortening in renal cell carcinoma. *Cancer Res.*, **54**. 236–41.

Morin, G. B. (1989). The human telomere terminal transferase enzyme is a ribonucleoprotein that synthesizes TTAGGG repeats. *Cell*, **59**, 521-9.

Olovnikov, A. M. (1971). Principle of marginotomy in template synthesis of polynucleotides. (In Russian). *Dokl. Akad. Nauk S.S.S.R.*, **201**, 1496-9.

Olovnikov, A. M. (1973). A theory of marginotomy. *J. Theor. Biol.*, **41**, 181-90.

Pathak, S., Wang, Z., Dhaliwal, M. K., and Sacks, P. C. (1988). Telomeric association: another characteristic of cancer chromosomes? *Cytogenet. Cell Genet.*, **47**, 227-9.

Radna, R. L., Caton, Y., Jha, K. K., Kaplan, P., Li, G., Traganos, F., and Ozer, H. L. (1989). Growth of immortal simian virus 40 *ts*A-transformed human fibroblasts is temperature dependent. *Mol. Cell. Biol.*, **9**, 3093-6.

Saltman, D., Morgan, R., Cleary, M. L., and de Lange, T. (1993). Telomeric structure in cells with chromosome end associations. *Chromosoma*, **102**, 121-8.

Vaziri, H., Schächter, F., Uchida, I., Wei, L., Zhu, X., Effros, R., Cohen, D., and Harley, C. B. (1993). Loss of telomeric DNA during aging of normal and trisomy 21 human lymphocytes. *Am. J. Human Genet.*, **52**, 661-7.

Wright, W. E., Pereira-Smith, O. M., and Shay, J. W. (1989). Reversible cellular senescence: implications for immortalization of normal human diploid fibroblasts. *Mol. Cell. Biol.*, **9**, 3088-92.

Yu, G.-L, Bradley, J. D., Attardi, L. D., and Blackburn, E. H. (1990). *In vivo* alteration of telomere sequences and senescence caused by mutated *Tetrahymena* telomerase RNAs. *Nature*, **344**, 126-32.

8

Chromatin structure and position effects

In *Saccharomyces cerevisiae* telomeric DNA sequences can influence the expression of genes located close to them, and may also influence the timing of replication of adjacent genomic regions. Genetic analysis has revealed that this position effect on gene expression is closely related to the transcriptional repression that occurs at the silent mating-type loci. Because the latter has been characterized in great detail it will be described first, thus enabling telomere position effects to be reviewed in context.

Sex and the single cell

Saccharomyces cerevisiae has three phenotypically distinct cell types. Haploid cells can be one of two mating types, **a** and α. Haploid cells of opposite mating types can mate to form $\mathbf{a}/\alpha$ diploids; these are unable to mate but can undergo meiosis and sporulation. Mating type is determined by the DNA sequence at the mating-type locus, *MAT*; **a** cells have the **a**1 and **a**2 genes at *MAT*, whereas α cells have the α1 and α2 genes. Yeast mating type has been the subject of many reviews, for example Haber (1992).

Haploid cells which express the HO endonuclease are able to switch mating type. Switching from α to **a** mating type is the result of replacing α-specific sequences at *MAT*, the α1 and α2 genes, with **a**-specific sequences, the **a**1 and **a**2 genes; switching from **a** to α mating type involves the converse sequence alteration. This replacement is achieved by a gene conversion event initiated by a double-strand break at *MAT* caused by the HO endonuclease. The donor of the sequences in the gene conversion event is one of two transcriptionally silent loci, *HML* and *HMR*. In most strains copies of the α1/α2 genes are found at the *HML* locus and copies of a1/a2 at HMR.

The a1/a2 and α1/α2 genes at the silent mating-type loci *HMR* and *HML* contain all the DNA sequences necessary for their expression yet both loci are transcriptionally inactive. Furthermore, only *MAT* is cleaved by HO, despite the *HM* loci also having the target sequence for HO cleavage. *HML* and *HMR* are flanked by *cis*-acting sequences termed 'silencers' which confer transcriptional repression. These act to promote the establishment of an altered chromatin structure at the *HM* loci which both represses transcription and prevents access of HO endonuclease.

The phenotype of silencing at the *HM* loci

Silencing of the inactive mating-type loci has been extensively reviewed (Laurenson and Rine 1992). The *HM* silencers are able to repress a variety of genes placed close to them, including ones transcribed by RNA polymerase II and ones transcribed by RNA polymerase III. Silencer orientation is unimportant and its distance from the gene does not need to be precise. Because a number of different genes are repressed by adjacent silencer elements and because these regions are inaccessible to the HO endonuclease *in vivo* this suggests that there is a general inaccessibility of the DNA template to a variety of proteins such as transcription factors, polymerases, and nucleases.

A number of *trans*-acting factors involved in mating-type silencing have been identified. Mutations in the *SIR2, SIR3,* and *SIR4* (silent information regulator) genes lead to complete derepression of both *HML* and *HMR*. Mutations in *SIR1* appear to lead to partial derepression of both loci (see below). None of the *SIR* genes are essential and, although all four have been cloned and sequenced, relatively little is known about the proteins they encode.

A *sir1* mutation appears to give partial derepression when judged at a population level. However, the culture is in fact a mixture of expressing and repressed cells, as demonstrated by assaying the phenotype of individual cells (Pillus and Rine 1989). A cell which is phenotypically **a** responds to the α-factor mating pheromone, which is produced by α cells, by undergoing cell cycle arrest and visibly changing its cellular morphology into what is termed a 'shmoo' in preparation for mating. Cells which are phenotypically **a**/α do not respond to α-factor in this way. Derepression of *HMLα* in a *MAT***a** strain causes the cell to adopt an **a**/α phenotype and it thus no longer responds to α-factor. By observing cells growing in the presence of α-factor it is possible to determine the *HMLα* expression state of individual cells. These studies revealed that the apparent partial derepression of *HMLα* in *sir1* strains actually reflects a mixed population of cells with *HMLα* either expressing or repressed (Pillus and Rine 1989). Furthermore, pedigree analysis demonstrated that both of the two transcriptional states were stable for many mitotic cell divisions and only occasionally switched; the progeny of a repressed cell were primarily repressed and the progeny of a derepressed cell were primarily derepressed. This is in contrast to wild-type cells, which rapidly re-established the repressed state (Miller and Nasmyth 1984). It was therefore suggested that the role of *SIR1* is to promote the establishment of the repressed state and that it is not involved in its maintenance from one generation to the next.

It is important to distinguish the metastable repression caused by *HMLα* in a *sir1* strain from the phenotype that can be caused by low-level expression, because this can at first sight give a similar effect. This has been illustrated by expressing the $\alpha 2$ gene in a *MATa* strain using the weak *GAL10* promoter (Mahoney *et al.* 1991). Expression of the $\alpha 2$ gene is sufficient to make a *MATa* cell lose responsiveness to α-factor. On media producing low level of expression of the *GAL10-$\alpha 2$* construct the cell population consists of a mixture of α-factor sensitive and resistant cells, which at first appears similar to the effect of a *sir1* mutation on *HMLα* expression. However, pedigree analysis reveals that the transcriptional state is very unstable and switches in both directions at a high frequency. Mahoney *et al.* (1991) suggest that in the cell population there is stochastic variation in the amount of the $\alpha 2$ gene product, and that the threshold at which a cell switches from being phenotypically α-factor sensitive to resistant falls within the continuum of $\alpha 2$ expression levels. The result is that, despite a continuous distribution of expression levels, the population behaves as if it consists of two cell types. The metastable repression at *HMLα* in a *sir1* strain is the result of a different process; the population is a mixture of cells either fully expressing the gene or fully repressed, rather than a continuum of gene expression levels. Although both scenarios produce a mixed population of cells as judged phenotypically, the most obvious difference is that low level expression produces an unstable phenotype, whereas *HMLα* is stably maintained in either the repressed or expressing state for many cell generations in a *sir1* strain.

The role of RAP1, ORC, and ABF1 in *HM* silencing

The *HMR* silencer element termed *HMR* E has been extensively analysed by both genetic and biochemical approaches. It contains binding sites for two abundant DNA binding proteins, RAP1 (repressor/activator protein) and ABF1 (ARS-binding factor). Both genes are essential for viability and act as transcriptional activators of a number of yeast genes unrelated to mating type. In contrast, both have roles in a transcription repression system at *HMR*. ABF1 is implicated in *HM* silencing both by deletion studies which remove the binding site (Laurenson and Rine 1992) and by the observation that mutations in *ABF1* can interfere with *HM* silencing (J. Rine, personal communication). There is no evidence that SIR1, 2, 3, or 4 bind directly to either of the *HM* silencers, and it is speculated that these are recruited to the silencers by protein–protein interactions with silencer binding factors such as RAP1, ORC, or ABF1.

RAP1 is implicated in *HM* silencing both by deletions which remove the binding site and mutations in the RAP1 gene. In general such experiments are complicated by functional redundancy at *HMR* E, in that removal of the ORC, ABF1, or RAP1 binding site alone has little effect, but removing any two severely affects silencing (Laurenson and Rine 1992). So although various *rap1* mutations can affect *HM* silencing, they only have a marked effect in combination with a partial silencer deletion. Mutations in *RIF1*

(RAP1-interacting factor; see Chapter 5) can also cause derepression in combination with *HMR* silencer mutations. The use of an *ADE2* gene under the influence of the *HMR* E silencer provides a colony sectoring assay which can measure both repression and the mitotic stability of the repressed or expressing state (Sussel *et al.* 1993). If the *ADE2* gene is expressed in the appropriate genetic background the yeast colony is white in colour on plates containing limiting amounts of adenine. However, if the gene is repressed the adenine biosynthetic pathway is blocked and an adenine precursor accumulates which produces a red colony colour. These strains with *ADE2* under the influence of *HMR* E produce colonies with sectors of red and white, indicating that the *ADE2* gene can be stably maintained in expressing and repressed states for a considerable number of cell divisions. Using this system the phenotype of *rap1^s* mutations, which specifically affect silencing at the *HM* loci, suggests that RAP1 has a role in the establishment rather than maintenance of the repressed state (Sussel *et al.* 1993), similar to the hypothesized role of SIR1 at *HMLα* (above).

Another element implicated by deletion studies in *HM* silencer function is an ARS element. Such elements provide origins of replication for yeast plasmids and can also function as chromosomal origins of replication. A multi-subunit protein complex, ORC, has been identified which binds to ARS elements *in vitro* (Bell and Stillman 1992). Mutations in the gene encoding one component (*ORC2*) can interfere with *HM* silencer function (Bell *et al.* 1993; Foss *et al.* 1993; Micklem *et al.* 1993; reviewed by Newlon 1993). Mutations in the gene encoding another subunit (*ORC5*) can also affect silencing (cited in Foss *et al.* 1993). Miller and Nasmyth (1984) derepressed *HMR* (using a *sir3^{ts}* allele at the restrictive temperature) and then determined if silencing could be re-established in the presence of various cell cycle blocks following a return to the permissive temperature. Their data suggested that passage through S phase was necessary to re-establish transcriptional silencing at *HMR*. However, it remains unclear whether initiation of DNA replication at the silencer ARS is necessary, or some other event involving ORC which occurs only in S phase. Indeed, although direct analysis has shown that *HMR* silencer acts as an origin of replication *in vivo*, *HML* does not appear to act as an origin in most cell cycles (Dubey *et al.* 1991; Rivier and Rine 1992). Another link between S phase and silencing is the observation that some *cdc7* mutations are able to suppress certain sequence mutations in the *HMR* E silencer which abolish repression, and that overexpression of *CDC7* interferes with silencing at the silent mating-type loci (Axelrod and Rine 1991). *CDC7* encodes a protein kinase which appears to be necessary for the G_1 to S transition and the initiation of mitotic DNA replication; its role in silencing is not understood.

Telomeric position effects

Yeast telomeres behave in a fashion similar to *HM* silencers, in that a number of different genes have been shown to be repressed when located adjacent to telomeric TG_{1-3} tracts (Gottschling *et al.* 1990). As with *HML* repression in *sir1* cells (Pillus and Rine 1989) it is possible to obtain a population containing a mixture of expressing and non-expressing cells with telomerically located genes. Furthermore, the repressed and expressing states are stable for many mitotic generations, as shown either using the *ADE2* sectoring assay or by pedigree analysis of cells with a telomeric *URA3* gene. In the latter, cells expressing *URA3* are able to grow on plates lacking uracil, whereas cells where *URA3* is repressed are able to grow in the presence of FOA. Using this assay it is found that repressed cells generally give rise to repressed progeny and vice versa.

Mutations in the *SIR2, SIR3, SIR4, NAT1, ARD1,* and *HHF2* (histone H4) genes abolish repression at telomeres as well as interfering with *HM* silencing (see below). Consistent with the instability of telomeric repression, which is similar to the instability of *HML* repression in a *sir1* strain (Pillus and Rine 1989), mutations in *SIR1* have no effect on telomere position effect (Aparicio *et al.* 1991).

The degree of repression is affected by both the promoter's strength and its distance from the telomere (Gottschling *et al.* 1990; Renauld *et al.* 1993). Although TG_{1-3} repeats are sufficient to cause *SIR1*-independent telomeric repression, the type of DNA separating the affected gene and the telomere can also affect the degree of repression. For example, the yeast subtelomeric Y' element appears to cause the repression to spread much further than normal. It is possible to speculate that this is because Y' elements contain both ABF1 and ORC binding sites close to the junction with the terminal TG_{1-3} tract. As similar binding sites function together with RAP1 binding sites to produce silencing at the *HM* loci, it is possible that the ABF1 and ORC binding sites in the Y' element act together with the RAP1 bound to the terminal TG_{1-3} repeats to form a more efficient silencer, although this has yet to be tested.

Telomeric position effects are very sensitive to the length of the TG_{1-3} array. Kyrion *et al.* (1993) have studied the effect of *rap1^t* alleles on telomeric position effects. These mutations produce a C-terminal truncated RAP1 protein which is still capable of binding DNA. They abolish telomeric position effects but also have a dramatic effect on telomere length. Terminal TG_{1-3} tract length increases from the wild-type (about 300 bp) to a range of sizes from 300 bp to over 4000 bp in *rap1^t* strains. The length is also very unstable, with long tracts being capable of rapid deletion of up to 3000 bp in a single generation (Kyrion *et al.* 1992). As Renauld *et al.* (1993) have shown that position effect decreases exponentially as a gene is placed further away from a telomere, is the loss of repression in *rap1^t*

strains because the elongated TG_{1-3} tracts cause the end of the chromosome to be further away from the reporter gene? This does not appear to be the case, as a telomere with a long TG_{1-3} tract can be segregated from a *rap1'* strain into a wild-type background and a *URA3* gene adjacent to such an elongated telomere is still subject to repression. In fact, a wild-type strain with elongated telomeres actually shows increased repression of telomere-adjacent sequences. A 4-fold increase in the length of the TG_{1-3} tract next to a *URA3* gene can produce a 500-fold increase in the frequency of repressed (FOA-resistant) colonies (Kyrion *et al.* 1993). Therefore the derepression in *rap1'* strains is not an indirect effect because of a change in telomere length.

In summary, the genetic evidence suggests that *HM* and telomeric repression are two variants on the same theme. One possibility is that the silent mating-type loci may have extra *cis*-acting sequences which in concert with SIR1 produce stable repression at the *HM* loci. These *cis*-acting sequences may be missing from telomeric TG_{1-3} arrays and thus telomeric silencing is both *SIR1*-independent and unstable. This speculation is discussed in more detail in a later section.

Altered chromatin structure at the silent mating-type loci and telomeres

The observation that *HM* silencing is to a large extent insensitive to promoter type, distance, and orientation suggests an altered chromatin structure in these regions making the DNA template generally inaccessible to a range of proteins including polymerases and transcription factors. This view is strengthened by a number of both *in vitro* and *in vivo* chromatin studies. Nasmyth (1982) has shown that chromatin at the silent mating-type loci is less accessible *in vitro* to DNase I than the equivalent sequences at *MAT* and that this inaccessibility is dependent on the *SIR* genes. Because a promoter deletion which renders the *HM* locus transcriptionally inactive even in a Sir⁻ strain still shows a *SIR*-dependent change in nuclease accessibility this suggests that the increase in accessibility to *HM* silencers in a Sir⁻ strain is not the result of transcription *per se*.

In vivo only the *MAT* locus is cut by the HO endonuclease, but in a Sir⁻ strain the silent mating-type loci are also cut (Strathern *et al.* 1982; Klar *et al.* 1984). The accessibility of the DNA template to another protein has also been tested *in vivo*, namely *E. coli dam* methyltransferase. This has been expressed in *S. cerevisiae*, and methylation at various sites in the genome can then be tested by the use of methylation-sensitive restriction enzymes. Both the silent mating-type loci and telomere-adjacent sequences show reduced accessibility to *dam* methyltransferase in a *SIR*-dependent fashion (Gottschling 1992; Singh and Klar 1992). The abolition of telomeric position effect by a *rap1'* mutation also correlates with an increased accessibility to *dam* methyltransferase (Kyrion *et al.* 1993). All these results suggest that the silent mating-type loci and telomeres adopt an altered chromatin structure which causes a decrease in accessibility to the DNA template.

What is the role of silencing at the mating-type loci? The obvious requirement is to repress the copies of the genes of the opposite mating-type, otherwise the cell would be phenotypically a/α. The reason such a specialized repression system is used may be that it also prevents access of HO endonuclease to the silent mating-type loci. HO cleavage at *MAT* initiates gene conversion, defining *MAT* as the acceptor site and the silent loci as donors. This provides a directionality in the gene conversion event; without it an *HMLα MATα HMRa* strain could gene convert *HMRa* and lose all copies of a-specific information. Assembling the silent mating-type loci into a silenced chromatin structure achieves the dual roles of transcriptional repression of the *HM* loci and maintaining directionality by protecting the *HM* loci from HO cleavage.

The biochemical nature of silenced chromatin

Another link between the observed position effect at telomeres and silencing at the *HM* loci is the biochemical nature of the repressive chromatin. Active versus inactive chromatin is a field that has been the subject of a large number of reviews (e.g. Felsenfeld 1992; Kornberg and Lorch 1992) and a full description is beyond the scope of this chapter. The nature of the 'silent chromatin' at *HMR/HML* is only partially understood but the *in vitro* and *in vivo* inaccessibility of the silent mating-type loci and telomeres to nucleases and methylases suggests that the transcriptional repression reflects a general inaccessibility of the chromatin to transcription factors and polymerases. This inaccessibility is a *SIR*-dependent function, suggesting the *SIR* genes act to assemble this altered chromatin structure. Biochemical and genetic evidence implicates histone acetylation in this altered chromatin structure.

The four core histones (H2A, H2B, H3, and H4) are among the most evolutionarily conserved eukaryotic proteins. The X-ray crystal structure of the nucleosome reveals the organization of the C-terminal globular domains of all four core histones, but none of the N-terminal regions are resolved. A biological role for these flexible N-terminal tails is inferred from the observation that their amino acid sequence is as highly conserved as that of the globular domains. It was therefore somewhat of a surprise to discover that a yeast strain expressing a histone H4 gene with an N-terminal deletion as the only source of histone H4 is viable (reviewed by Smith 1991).

What these strains do show is a derepression of the silent mating-type loci. The N-terminal tails of the core histones are rich in basic amino acids and are subject to a range of post-translational modifications, including methylation, acetylation, and phosphorylation. Four lysine residues are found in the first 16 amino acids of histone H4; acetylation of their ε-groups neutralizes their positive charge. Point mutations in histone H4 have been assayed for their effect on silencing (Park and Szostak 1990; Johnson *et al.* 1992; reviewed by Laurenson and Rine 1992). Single changes in any of the first three lysines have little effect on silencing. However, although mutations that substitute a positively charged residue for lysine 16 have little effect, mutations that substitute an uncharged residue are derepressed. This suggested that reduced acetylation of lysine 16 might be associated with silencing.

Histone acetylation is associated with active chromatin in a number of systems (reviewed by Turner 1993). For example, nucleosomes with H4 acetylated at lysine 16 are specifically associated with the hyperexpressed X chromosome in male *Drosophila*. Similarly, the inactive X chromosome in female mammals is depleted for acetylated histone H4 (Jeppesen and Turner 1993). In *Drosophila*, mutations in the *Su-var(2)1* gene reduce the repression of gene expression caused by adjacent heterochromatin in flies exhibiting position effect variegation (Dorn *et al.* 1986). Such flies show increased histone acetylation, which again suggests a link between deacetylation and repression of gene expression. In yeast there is now direct biochemical evidence from immunoprecipitation experiments that silencing at *HML/HMR* is associated with underacetylation of histone H4 (Braunstein *et al.* 1993). This underacetylation requires the products of the *SIR2, SIR3,* and *SIR4* genes. The data also suggest that the telomeric Y' elements are somewhat depleted for acetylated histone H4. Overexpression of *SIR2* (but not *SIR3*) causes a decrease in the acetylation of histones H2B, H3, and H4, and it is possible to speculate that this suggests that the biochemical function of SIR2 *in vivo* is to deacetylate these three core histones, although this has not been shown directly. The depletion of acetylated histone H4 is consistent with a similar repressive chromatin structure being assembled at both telomeres and the silent mating-type loci.

The mechanism by which histone acetylation affects transcription is not clear. One possibility is that it directly interferes with transcription factor access to the DNA template. There is evidence in support of this hypothesis from *in vitro* studies using the *Xenopus* 5S rRNA gene and transcription factor TFIIIA (Lee *et al.* 1993). TFIIIA can form a tertiary complex with the 5S rRNA gene associated with acetylated histones but cannot associate with the gene if it is wound round a non-acetylated core particle. This suggests that histone acetylation, in itself, is sufficient to affect accessibility of proteins to the DNA template. One model is that the N-terminal tails of the core histones lie in the major groove of the DNA helix wound around the core particle. If they are not acetylated this interaction is strong and stable, and steric hindrance caused by the histone tails prevents access of other proteins to the DNA. Acetylation could reduce the strength of this interaction and thus allow access to the DNA. It is therefore possible that the inaccessibility of the DNA in silent chromatin in yeast to a range of proteins such as nucleases and methylases is a similar direct consequence of histone deacetylation. One caveat is that histone H4 mutations have a marked effect on *HML* but only a weak effect on *HMR*. This suggests that deacetylation of histone H4 is not sufficient in itself for silencing at *HMR*. Although N-terminal deletions of H2A, H2B, and H3 have little effect on *HMR* silencing, the possibility that silencing at *HMR* involves the deacetylation of more than one core histone is difficult to rule out because these histone deletions cannot be tested in combination with H4 deletions; such strains are non-viable. If *HMR* has redundancy in its silencing mechanism as compared with *HML* this would provide one explanation why various mutations (*nat1*, *hhf2*, and *ard1*) have a more marked effect on *HML*.

Mutations in two other genes, *NAT1* and *ARD1*, interfere with silencing at *HML*. These genes encode an N-terminal acetyltransferase which acetylates the N-terminal residue of at least 30 different proteins (Park and Szostak 1992). It should be emphasized that the acetylation they cause is not the same as the post-translational lysine ε-group acetylation of histone H4 found in silenced chromatin. The acetylation caused by NAT1/ARD1 is irreversible, cotranslational, and specific for the α-amino group of a wide variety of proteins, and is thought to interfere with protein stability rather than to alter gene expression. It is not known which is the cellular substrate responsible for the effect of *nat1/ard1* mutations on silencing. Analysis of histone H4 variants indicates that this is not the biologically significant substrate (Park and Szostak 1990; Laurenson and Rine 1992) and probably neither are the three other core histones, as N-terminal deletions of them have little effect on *HML* silencing. *ARD1/NAT1* may affect the stability of one or more of the other proteins involved in silencing.

A model for telomeric position effect

There are no mutations that affect telomere position effect which have not also been shown to interfere with some aspect of *HM* silencing. Even the *rap1*[t] mutations, which have such a dramatic effect on telomeric position effect, also affect silencing at *HML* (Kyrion *et al.* 1993). In contrast, there are a number of mutations which affect silencing but not telomere position effect, such as *sir1*, *rif1*, and *rap1*[s] (Sussel and Shore 1991; Hardy *et al.* 1992). This argues that telomere position effect reflects a subset of the activities which produce silencing at the *HM* loci, possibly reflecting that a telomeric silencer lacks some of the *cis*-acting sequences (such as ORC and ABF1 binding sites) which are found at *HM* silencers. Since some *HM* repression can be exerted in a *sir1* strain there must be a *SIR1*-independent pathway of silencing, and as telomeric silencing is *SIR1*-independent it is possible that it is this subset of *HM* silencing that is being observed at telomeres.

In contrast to the single RAP1 binding site at *HMR*, there are about 10–20 RAP1 binding sites in a TG_{1-3} repeat array of wild-type length (Gilson *et al.* 1993). Can such a large array of RAP1 binding sites establish silencing in the absence of the other sequence elements found at *HMR*? There is evidence from a two-hybrid assay that RAP1 can interact with both SIR3 and SIR4 *in vivo* (D. Shore, personal communication). This inter-action may provide a way for a large array of RAP1 proteins to act as a silencer. Why does this form of silencing seem to be specific to terminal TG_{1-3} tracts? This question has not been rigorously addressed. A 300 bp terminal tract causes repression but an 81 bp internal tract does not (Gottschling *et al.* 1990). Telomeric repression still occurs in a *tel1* strain even though the average telomere length is 120 bp. However, as telomeric silencing is very sensitive to the length of the TG_{1-3} array, it may be that the internal TG_{1-3} tracts tested have not been long enough to promote silencing. It would be premature to argue, for example, that telomeric silencing reflects a protein which binds terminal but not internal TG_{1-3} tracts, such as an end-binding protein. Although the reason for the apparent specificity of silencing to terminal TG_{1-3} tracts is unknown it is possible to speculate that it reflects a straightforward steric problem. It is known that terminal TG_{1-3} tracts in chromatin are protected against nuclease digestion yet are non-nucleosomal (Wright *et al.* 1992). RAP1 is a major component of this terminal 'telosome' and protection against nuclease digestion may reflect the large number of RAP1 molecules bound, as similar protection is seen using RAP1 and naked DNA *in vitro* (Gilson *et al.* 1993). Assembly of such an array of RAP1 proteins is likely to require nucleosome displacement. Nucleosomes can slide along the template, but to displace a nucleosome from an internal TG_{1-3} tract in this fashion would also require any adjacent nucleosomes to move as well. In contrast, there are by definition no nucleosomes at the end of a chromosome after the most distal one, which may facilitate the displacement of the nucleosome over a terminal TG_{1-3} tract. If a silencer is composed of a large array of bound RAP1 then this simple boundary effect may make it easier to assemble such an array at a terminal location.

It has been speculated that one important function of the *HM* silencer is to recruit SIR1. This model is based on the observation that the binding site for the GAL4 protein can act as a silencer in a strain expressing a hybrid protein consisting of SIR1 fused to the DNA binding domain of GAL4 (Chien *et al.* 1993). It is speculated that SIR1 recruitment normally occurs by protein–protein interactions with silencer binding factors such as ORC or ABF1, although there is no direct evidence for such an interaction as yet. It is further speculated that telomeric silencers are *SIR1*-independent because they lack ORC and ABF1 sites, although again there is no direct evidence for this. In strains expressing the SIR1–GAL4 fusion protein (Chien *et al.* 1993) a GAL4 binding site adjacent to a telomeric TG_{1-3} tract dramatically improves the silencing of an adjacent *URA3* gene, so much so that the frequency of repressed (FOA-resistant) cells is almost identical to that seen in a *ura3⁻* strain. This effect is dependent upon the expression of the SIR1–GAL4 fusion protein. It is therefore possible that telomeric position effect is unstable and *SIR1*-independent for the same reason, that is a lack at telomeres of the sequences which tether SIR1 to the *HM* silencers.

Position effect variegation in *Drosophila*

The phenomenon of position effect variegation at yeast telomeres is of interest because of its similarity to classical position effect variegation in *Drosophila*, which has been studied for many decades (Henikoff 1992). *Drosophila* position effect variegation is usually the result of a genomic rearrangement which relocates a gene to a position close to a region of heterochromatin, such as is found at *Drosophila* centromeres (see Chapter 9). Repressive chromatin is believed to spread from the heterochromatic domain to the variegated gene in a stochastic fashion, leading to repression of expression. Variations in the extent of spreading lead to mosaicism, producing a mixed population of cells either expressing or not expressing the gene. Indeed, cell to cell differences in the extent of the spread of the heterochromatin can be seen cytologically; in some cells the variegated gene is visibly heterochromatic, in other cells not. The other hallmark of this system is that the repressed or expressing state is stable for a number of mitotic divisions. In the case of the white *gene*, which affects eye colour, this leads to patches of expressing tissue interspersed with patches of non-expressing tissue.

Telomere position effect in yeast has the two hallmarks of *Drosophila* position effect: a mixed population of expressing and repressed cells and only occasional switches between the two transcriptional states. This mitotic stability is clearly shown using an *ADE2* sectoring assay. There has been considerable success in using position effect variegation in *Drosophila* as an assay to identify proteins involved in heterochromatin formation (Reuter and Spierer 1992). Genetic analysis has identified a number of genes which affect position effect variegation. *Su(var)* mutations suppress the inactivation of the variegated gene, whereas *En(var)* mutations enhance the inactivation. *Su(var)205* encodes heterochromatin protein 1 (HP1), which is predominantly associated with *β*-heterochromatin in polytene chromosomes and has now been found to have homologues in a number of other species, including mammals (Henikoff 1992: Shaffer *et al.* 1993). Although the altered chromatin structure at yeast telomeres is very similar to that found at the silent mating-type loci, position effect variegation at yeast telomeres may provide an additional genetic assay to those based on *HM* function to isolate components of the altered chromatin found at telomeres and the silent mating-type loci.

The *in vivo* role of telomeric silencing

The existence of position effect of telomeric genes in yeast does not necessarily indicate that the ability to repress adjacent genes in the genome is its *raison d'être*. Indeed, although there are genes located near yeast telomeres, such as members of the *PHO, MAL,* and *SUC* families (Carlson *et al.* 1985; Charron and Michels 1988; Charron *et al.* 1989; Venter and Hörz 1989), there is no evidence that they are subject to telomeric repression.

An essential telomere function is to make telomeric DNA inaccessible to proteins such as exonucleases, ligases, and recombination enzymes, as this is what makes a natural chromosome end different from a double-strand break. Silencing does not appear necessary to protect yeast telomeres; Sir$^-$ strains are viable, suggesting that telomere integrity is not being dramatically compromised, and neither *sir3Δ* nor *sir4Δ* mutations have a marked effect on the rate of loss of a non-essential chromosome (Palladino *et al.* 1993). In contrast, a non-essential yeast chromosome terminating in a double-strand break (for example produced by HO endonuclease cleavage *in vivo*) is rapidly lost (Weiffenbach and Haber 1981; Klar *et al.* 1984; Bennett *et al.* 1993). This argues that chromosome stability largely does not require *SIR*-dependent altered chromatin structure.

It is known that there are numerous binding sites for RAP1 in the terminal TG$_{1-3}$ arrays (Chapter 5). *In vitro* binding of RAP1 to naked TG$_{1-3}$ DNA is sufficient to protect stretches as large as 100 bp from nuclease digestion (Gilson *et al.* 1993). *In vivo* the terminal TG$_{1-3}$ repeats are assembled into a non-nucleosomal structure, perhaps reflecting RAP1 binding (Wright *et al.* 1992) and RAP1 binding may be sufficient to protect the end of the yeast chromosome *in vivo*. In contrast all the available data implicate alterations in nucleosomal chromatin structure in silencing. It is difficult to visualize how such altered chromatin could protect the end of the chromosome if nucleosomes do not extend that far. Instead, one might argue that silencing occurs in telomere-adjacent regions as a side-effect of using a large array of RAP1 proteins to protect the terminal TG$_{1-3}$ array.

Heterochromatin consists of those regions of the genome which are late-replicating and transcriptionally inactive, and remain condensed during interphase. In many species constitutive heterochromatin is composed of large tandem repeat arrays of a short sequence, typically less than 1 kb in length. In some species there are terminally located regions of heterochromatin; these have been detailed in Chapter 3. The species-specificity of many of these sequences may indicate that many do not have a positive role in cellular functions and have been perhaps amplified as a by-product of a process such as recombination. However, at least some heterochromatic sequences have been implicated in biological processes. For example, the mouse minor satellite may be involved in centromere function, possibly because it binds the centromere-associated CENP-B protein (Kipling *et al.* 1991). Heterochromatin has been suggested to be involved in chromosome pairing during distributive segregation in *Drosophila* meiosis (Hawley and Theurkauf 1993). The centromeric heterochromatin of mammalian chromosomes is the last region of the chromosome to separate prior to the metaphase–anaphase transition and defects in heterochromatin in *Drosophila* can cause premature sister-chromatid separation, thus interfering with mitotic chromosome segregation (Wines and Henikoff 1992). These various biological functions may require a region of condensed chromatin, and a side-effect of this might be that the expression of an adjacent gene is compromised. It is perhaps important to stress that although heterochromatin may confer position effects on adjacent genes, this is not necessarily its role *in vivo*; repression of transcription may simply reflect a general inaccessibility of the DNA template required for another reason.

Telomere position effects on replication

Eukaryotic chromosomes are replicated using numerous origins of replication along their length. This ensures that the genome is replicated in the time allocated to S phase, given the large size of the eukaryotic genome and the rate of replication fork progression. However, not all replication origins are 'fired' at the same point in S phase (reviewed by Fangman and Brewer 1992). Mammalian chromosomes are characterized by regional differences in replication timing; there are large regions which replicate late interspersed with earlier replicating regions. This can be shown cytologically using BrdU labelling of replicating DNA (Drouin *et al.* 1990). Each temporal domain contains on average 2.5 Mb of DNA, enough to encompass 10–50 replication origins, although regions as large as 10 Mb replicate during discrete intervals of S phase.

What determines the timing of replication? Heterochromatin is classically defined as late-replicating, but it is not possible to determine if constitutive heterochromatin replicates late because of its chromatin structure or because it contains *cis*-acting sequences which determine its replication timing. However, in female mammals the active X chromosome replicates earlier than the transcriptionally inactive X, demonstrating an epigenetic regulation of replication timing. The mechanism underlying this difference in replication timing is not known.

In yeast, chromosomal context appears to influence the time at which an origin of replication fires. *ARS501* is an origin of replication located about 20 kb from the telomere of chromosome V. Using two-dimensional gel electrophoresis techniques (reviewed by Fangman and Brewer 1991) it has been shown that this sequence is an origin of chromosomal replication (Ferguson *et al.* 1991). Using Meselson–Stahl density-transfer experiments it is possible to ask at what point during S phase any yeast sequence is replicated. The timing of sequences on chromosome V around *ARS501* supports its designation as an origin, as sequences flanking it on both sides replicate later in S phase, as if by replication forks emanating from *ARS501*. *ARS501* fires much later in S phase than another well-characterized yeast origin, *ARS1*. Is this because the sequence of the ARS itself determines the timing of replication, perhaps because it interacts with different *trans*-acting factors? This does not appear to be the case, because if *ARS1* is moved to a site immediately adjacent to *ARS501* on chromosome V it no longer replicates early in S phase.

ARS501 is located close to a telomere. As subtelomeric Y′ elements replicate late in S phase (McCarroll and Fangman 1988) it was tested whether location close to a telomere affects replication timing (Ferguson and Fangman 1992). A 15 kb region of chromosome V encompassing *ARS501* cloned into a circular yeast plasmid replicates early in S phase, yet an identical construct in a linear form, with TG_{1-3} repeats added to the ends, replicates late in S phase. In similar experiments Wellinger *et al.* (1993) use two-dimensional gel electrophoresis techniques to show that the 2μm ARS fires early on a circular plasmid yet late on a linear vector. Together these experiments suggest that the timing of firing of an origin is not determined by its sequence but the chromosomal context in which it finds itself, and that one feature which influences timing is presence close to a telomeric TG_{1-3} repeat array. One prediction of this model is that the ARS elements within the subtelomeric Y′ elements would fire late in S phase. This has been confirmed by two-dimensional gel analysis which demonstrates that the Y′ ARS is functional as an origin (Ferguson *et al.* 1991) and density-transfer experiments which show that Y′ elements replicate late in S phase (McCarroll and Fangman 1988). In separate experiments on origin usage on chromosome III there is preliminary evidence to suggest that in some cases locating an ARS close to a telomere can abolish origin function completely (Newlon *et al.* 1993).

It is not known if this late replication has any relationship to the *SIR*-dependent telomeric position effects on transcription. It is possible that the repressive chromatin emanating from the telomere is able to influence the time of firing of an origin. Another possibility is that this effect reflects the compartmentalization of replication in the nucleus. In higher eukaryotes replicating DNA has been pulse-labelled and the distribution of replication in the S phase nucleus examined microscopically (Jackson 1990). For a mammalian cell this typically reveals replication to be confined to a series of small spots in the nucleus (O'Keefe *et al.* 1992; Hassan and Cook 1993). Using synchronized mouse fibroblasts and confocal microscopy Fox *et al.* (1991) have followed the pattern of replication through S phase. In early S phase replication occurs in 100–300 small foci throughout the nucleus, with the exception of the condensed heterochromatic regions. In mid to late S phase the spots are larger and confined to the condensed heterochromatic regions. Dramatically, at the end of S phase replication is confined to the region immediately underneath the nuclear envelope. This opens up the possibility that locating a sequence close to a yeast telomere, because of the localization of RAP1, results in it being redistributed to a region of the nucleus characterized by late replication.

Telomeres, silencers, and plasmid segregation: another associated phenotype

Saccharomyces cerevisiae circular plasmids which do not contain a centromere are mitotically unstable. They mis-segregate in mitosis, and when they do so there is a very strong bias towards the plasmid remaining in the mother cell as opposed to the bud (the daughter). There are a number of systems which overcome this instability that do not involve conventional centromere-based segregation. The first is shown by the endogenous yeast $2\,\mu m$ plasmid. Two plasmid-encoded proteins (REP1 and REP2) and a *cis*-acting sequence (*REP3*) are required for plasmid stability (Wu *et al.* 1987; reviewed by Farrar and Williams 1988). REP1 is a component of the yeast nuclear matrix and may be a DNA binding protein which interacts with *REP3*. Secondly, acentric plasmids containing the *HMR* E silencer display increased mitotic stability. The third example is provided by circular plasmids containing yeast TG_{1-3} repeats (Longtine *et al.* 1992, 1993; Enomoto *et al.* 1994).

As both REP1 and RAP1 are components of the nuclear matrix (see Chapter 5) one possibility is that segregation of these various acentric plasmids reflects binding of the plasmid to a nuclear matrix component which is equipartitioned to both progeny. However, a number of yeast ARS elements are matrix-associated yet do not show this increase in mitotic stability (Amati and Gasser 1988). Another possibility is that it reflects an abolition of the mother–daughter segregation bias. However, when plasmids containing TG_{1-3} repeats mis-segregate they still show the strong bias towards remaining in the mother cell. Furthermore, *Schizosaccharomyces pombe* telomere sequences confer a similar mitotic stability on circular plasmids in *S. pombe* (Longtine *et al.* 1992). This yeast divides by symmetrical fission and thus there is no mother-daughter bias to be eliminated. This argues that the mitotic stability conferred by TG_{1-3} repeats is not the result of abolishing a segregation bias but the result of reducing the total mount of mis-segregation.

The improved mitotic stability conferred by the *HMR* E silencer is totally abolished by *sir2*, *sir3*, and *sir4* mutations (Kimmerly and Rine 1987;

Kimmerly *et al.* 1988). In contrast, TG_{1-3}-mediated stability is only partially reduced by these mutations (Longtine *et al.* 1993). This indicates that TG_{1-3} repeats can promote at least some improvement in segregation even in the absence of the *SIR* gene products. This segregation can be affected by some *rap1*^{ts} mutations at the semi-permissive temperature, suggesting that RAP1 has a role in TG_{1-3} mediated segregation.

In *S. cerevisiae* the RAP1 protein is clustered into about eight foci in the interphase nucleus even though there are 32 telomeres in the cell (see Chapter 5). Using antibodies against RAP1 the majority of the signal is seen at the ends of meiotic chromosomes. This, together with the *in vitro* evidence that RAP1 can bind TG_{1-3} arrays, has been taken as evidence that telomeric DNA sequences are clustered in the interphase nucleus, although there is no direct evidence for this as yet. This RAP1 clustering is dependent on at least the *SIR3* and *SIR4* genes, as mutations in these two genes changes its distribution (Palladino *et al.* 1993). The number of RAP1 foci is increased in these mutants, which would be consistent with a disruption of telomere-telomere associations. Although speculative at this stage, it has been suggested that the stability of TG_{1-3}-containing plasmids is caused by a physical association of the TG_{1-3} sequences with similar repeats found at the end of endogenous linear chromosomes. In effect the circular plasmid could be 'hitch-hiking' during mitosis by associating with a chromosome with centromere function (Longtine *et al.* 1993). Although TG_{1-3} arrays may associate with endogenous telomeres better in a Sir$^+$ strain, it is speculated that the large number of bound RAP1 molecules enables internal TG_{1-3} arrays to associate with telomeres even in Sir$^-$ strains, perhaps because RAP1 molecules are able to interact with each other directly. The plasmid stability phenotype produced by TG_{1-3} arrays provides an experimental tool with which to identify *trans*-acting factors that might be involved in altered chromatin structure at telomeres and in the organization of telomeres in the interphase nucleus.

Summary

Do telomeres exert position effects in other species? To answer this question one must clearly define position effect variegation. Position effect variegation does not, for example, require the total abolition of expression of a gene in every cell in the population. The phenotype of position effect variegation in both yeast and *Drosophila* is the formation of a mixed population of cells, some expressing and others repressed. However, at a population level the variegated gene is detectably expressed. Therefore, the observation that genes in a number of species are very close to a telomere yet expressed (Greider 1992) is insufficient to prove they are not subject to position effect variegation. Such telomeric genes include surface antigen genes in trypanosomes and *Paramecium*, and almost all the genes in

Oxytricha as these are found on small linear molecules in the macronucleus (Chapter 6). Although position effect variegation can be caused by locating a gene close to a *Drosophila* telomere, this is likely to reflect the presence of telomeric heterochromatin because a *white* gene next to a non-heterochromatic terminus is not subject to position effects (see Chapter 9; Levis 1989).

Surface antigen gene expression in trypanosomes is one of the better candidates for another example of telomeric position effect variegation. The trypanosome genome contains many surface antigen genes, only one of which is expressed at any one time (Pays and Steinert 1988). By switching between the expression of various surface antigen genes the parasite is able to evade the host immune system. Many of the surface antigen genes are located at telomeric sites but are not expressed. At any one time there is a single active expression site, located close to a telomere. Although many switches in surface antigen gene expression involve DNA rearrangements which transfer genes to such an expression site, there are examples where a previously inactive telomeric gene is activated without any sequence change, this being termed expression site switching. It has been suggested (Van der Ploeg *et al*. 1992) that such epigenetic control could reflect a telomeric position effect. An epigenetic control mechanism is supported by the observation that inactive trypanosome telomeres have secondary base modifications, including β-D-glucosyl-hydroxymethyluracil, which are absent from active telomeres (Bernards *et al*. 1984; Gommers-Ampt *et al*. 1993).

One criterion for position effect variegation is the stable maintenance of two different expression states. This can be demonstrated by white colour patches in a normally red *Drosophila* eye or *ADE2* sectors in *S. cerevisiae*. Telomeric position effect variegation has been discovered in *S. pombe*, which is evolutionarily distant to budding yeast and has a very different terminal repeat sequence (E. Nimmo, personal communication). Telomeric position effects may therefore be a widespread phenomenon.

Are there position effects on replication in other eukaryotes? Cytologically the terminal bands of human chromosomes can replicate at different times in S phase (Drouin *et al*. 1990). Using a $(TTAGGG)_n$ probe to detect all telomeres simultaneously it has been shown by density-transfer experiments that this sequence replicates throughout S phase (Ten Hagen *et al*. 1990). Therefore human telomeres are not consistently late-replicating by either criterion, suggesting that the timing of replication of mammalian $(TTAGGG)_n$ arrays is determined primarily by that of the adjacent genomic region. The clearest evidence for an origin of replication in a normal mammalian cell is that within the β-globin gene cluster (Kitsberg *et al*. 1993). This functions as an origin of bidirectional replication in all cell types, but the β-globin locus is characterized by early replication in cells where it is expressed and late replication in cells where it is not. Together this indicates that firing of the β-globin origin is under epigenetic control related to its transcriptional state. Epigenetic control of replication origin activity may therefore be a widespread phenomenon.

References

Reviews

Fangman, W. L. and Brewer, B. J. (1991). Activation of replication origins within yeast chromosomes. *Annu. Rev. Cell Biol.*, **7**, 375–402.

Fangman, W. L. and Brewer, B. J. (1992). A question of time: replication origins of eukaryotic chromosomes. *Cell*, **71**, 363–6.

Farrar, N. A. and Williams, K. L. (1988). Nuclear plasmids in the simple eukaryotes *Saccharomyces cerevisiae* and *Dictyostelium discoideum*. *Trends Genet.*, **4**, 343–8.

Felsenfeld, G. (1992). Chromatin as an essential part of the transcriptional mechanism. *Nature*, **355**, 219–24.

Greider, C. W. (1992). Telomere chromatin and gene expression. *Curr. Biol.*, **2**, 62–4.

Haber, J. E. (1992). Mating-type gene switching in *Saccharomyces cerevisiae*. *Trends Genet.*, **8**, 446–52.

Hawley, R. S. and Theurkauf, W. E. (1993). Requiem for distributive segregation: achiasmate segregation in *Drosophila* females. *Trends Genet.*, **9**, 310–17.

Henikoff, S. (1992). Position effect and related phenomena. *Curr. Opin. Genet. Dev.*, **2**, 907–12.

Jackson, D. A. (1990). The organisation of replication centres in higher eukaryotes. *Bioessays*, **12**, 87–9.

Kornberg, R. D. and Lorch, Y. (1992). Chromatin structure and transcription. *Ann. Rev. Cell Biol.*, **8**, 563–87.

Laurenson, P. and Rine, J. (1992). Silencers, silencing, and heritable transcriptional states. *Microbiol. Rev.*, **56**, 543–60.

Newlon, C. S. (1993). Two jobs for the origin replication complex. *Science*, **262**, 1830–1.

Pays, E. and Steinert, M. (1988). Control of antigen expression in African trypanosomes. *Ann. Rev. Genet.*, **22**, 107–26.

Reuter, G. and Spierer, P. (1992). Position effect variegation and chromatin proteins. *Bioessays*, **14**, 605–12.

Shaffer, C. D., Wallrath, L. L., and Elgin, S. C. R. (1993). Regulating genes by packaging domains: bits of heterochromatin in euchromatin? *Trends Genet.*, **9**, 35–37.

Smith, M. M. (1991). Histone structure and function. *Curr. Opin. Cell Biol.*, **3**, 429–37.

Turner, B. M. (1993). Decoding the nucleosome. *Cell*, **75**, 5–8.

Van der Ploeg, L. H. T., Gottesdiener, K., and Lee, M. G.-S. (1992). Antigenic variation in African trypanosomes. *Trends Genet.*, **8**, 452–57.

Primary papers

Amati, B. B. and Gasser, B. M. (1988). Chromosomal ARS and CEN elements bind specifically to the yeast nuclear scaffold. *Cell*, **54**, 967-8.

Aparicio, O. M., Billington, B. L., and Gottschling, D. E. (1991). Modifiers of position effect are shared between telomeric and silent mating-type loci in S. cerevisiae. *Cell*, **66**, 1279-87.

Axelrod, A. and Rine, J. (1991). A role for CDC7 in repression of transcription at the silent mating-type locus *HMR* in *Saccharomyces cerevisiae*. *Mol. Cell. Biol.*, **11**, 1080-91.

Bell, S. P. and Stillman, B. (1992). ATP-dependent recognition of eukaryotic origins of DNA replication by a multiprotein complex. *Nature*, **357**, 128-34.

Bell, S. P., Kobayashi, R., and Stillman, B. (1993). Yeast origin recognition complex functions in transcription silencing and DNA replication. *Science*, **262**, 1844-9.

Bennett, C. B., Lewis, A. L., Baldwin, K. K., and Resnick, M. A. (1993). Lethality induced by a single site-specific double-strand break in a dispensable yeast plasmid. *Proc. Natl. Acad. Sci. USA*, **90**, 5613-17.

Bernards, A., van Harten-Loosbroek, N., and Borst, P. (1984). Modification of telomeric DNA in *Trypanosoma brucei*; a role in antigenic variation? *Nucleic Acids Res.*, **12**, 4153-70.

Braunstein, M., Rose, A. B., Holmes, S. G., Allis, C. D., and Broach, J. R. (1993). Transcriptional silencing in yeast is associated with reduced nucleosome acetylation. *Genes Dev.*, **7**, 592-604.

Carlson, M., Celenza, J. L., and Eng, F. J. (1985). Evolution of the dispersed *SUC* gene family of *Saccharomyces* by rearrangements of chromosome telomeres. *Mol. Cell. Biol.*, **5**, 2894-902.

Charron, M. J. and Michels, C. A. (1988). The naturally occurring alleles of *MAL1* in *Saccharomyces* species evolved by various mutagenic processes including chromosomal rearrangements. *Genetics*, **120**, 83-93.

Charron, M. J., Read, E., Haut, S. R., and Michels, C. A. (1989). Molecular evolution of the telomere-associated *MAL* loci of Saccharomyces. *Genetics*, **122**, 307-16.

Chien, C.-t., Buck, S., Sternglanz, R., and Shore, D. (1993). Targeting of SIR1 protein establishes transcriptional silencing at *HM* loci and telomeres in yeast. *Cell*, **75**, 531-41.

Dorn, R., Heymann, S., Lindigkeit, R., and Reuter, G. (1986). Suppressor mutation of position-effect variegation in *Drosophila melanogaster* affecting chromatin properties. *Chromosoma*, **93**, 398-403.

Drouin, R., Lemieux, N., and Richer, C.-L. (1990). Analysis of DNA replication during S-phase by means of dynamic chromosome banding at high resolution. *Chromosoma*, **99**, 273-80.

Dubey, D. D., Davis, L. R., Greenfeder, S. A., Ong, L. Y., Zhu, J., Broach, J. R., *et al.* (1991). Evidence suggesting that the *ARS* elements associated with silencers of the yeast mating-type locus *HML* do not function as chromosomal DNA replication origins. *Mol. Cell. Biol.*, **11**, 5346-55.

Enomoto, S., Longtine, M. S., and Berman, J. (1994). Enhancement of telomere-plasmid segregation by the X-telomere associated sequence in *Saccharomyces cerevisiae* involves *SIR2, SIR3, SIR4* and *ABF1*. *Genetics*, **136**, 757-67.

Ferguson, B. M. and Fangman, W. L. (1992). A position effect on the time of replication origin activation in yeast. *Cell*, **68**, 333-39.

Ferguson, B. M., Brewer, B. J., Reynolds, A. E. and Fangman, W. L. (1991). A yeast origin of replication is activated late in S phase. *Cell*, **65**, 507–15.

Foss, M., McNally, F. J., Laurenson, P., and Rine, J. (1993). Origin recognition complex (ORC) in transcriptional silencing and DNA replication in *S. cerevisiae*. *Science*, **262**, 1838–44.

Fox, M. H., Arndt-Jovin, D. J., Jovin, T. M., Baumann, P. H., and Robert-Nicoud, M. (1991). Spatial and temporal distribution of DNA replication sites localized by immunofluorescence and confocal microscopy in mouse fibroblasts. *J. Cell Sci.*, **99**, 247–53.

Gilson, E., Roberge, M., Giraldo, R., Rhodes, D., and Gasser, S. M. (1993). Distortion of the DNA double helix by RAP1 at silencers and multiple telomeric binding sites. *J. Mol. Biol.*, **231**, 293–310.

Gommers-Ampt, J. H., Van Leeuwen, F., de Beer, A. L. J., Vliegenthart, J. F. G., Dizdaroglu, M., Kowalak, J. A., *et al.* (1993). β-D-glucosyl-hydroxymethyluracil: a novel modified base present in the DNA of the parasitic protozoan T. brucei. *Cell*, **75**, 1129–36.

Gottschling, D. E. (1992). Telomere-proximal DNA in *Saccharomyces cerevisiae* is refractory to methyltransferase activity *in vivo*. *Proc. Natl. Acad. Sci. USA*, **89**, 4062–5.

Gottschling, D. E., Aparicio, O. M., Billington, B. L., and Zakian, V. A. (1990). Position effect at S. cerevisiae telomeres: reversible repression of pol II transcription. *Cell*, **63**, 751–62.

Hardy, C. F. J., Sussel, L., and Shore, D. (1992). A RAP1-interacting protein involved in transcriptional silencing and telomere length regulation. *Genes Dev.*, **6**, 801–14.

Hassan, A. B. and Cook, P. R. (1993). Visualization of replication sites in unfixed human cells. *J. Cell Sci.*, **105**, 541–50.

Jeppesen, P. and Turner, B. M. (1993). The inactive X chromosome in female mammals is distinguished by a lack of histone H4 acetylation, a cytogenetic marker for gene expression. *Cell*, **74**, 281–9.

Johnson, L. M., Fisher-Adams, G., and Grunstein, M. (1992). Identification of a non-basic domain in the histone H4 N-terminus required for repression of the yeast silent mating loci. *EMBO J.*, **11**, 2201–9.

Kimmerly, W. J. and Rine, J. (1987). Replication and segregation of plasmids containing *cis*-acting regulatory sites of silent mating-type genes in *Saccharomyces cerevisiae* are controlled by the *SIR* genes. *Mol. Cell. Biol.*, **7**, 4225–37.

Kimmerly, W., Buchman, A., Kornberg, R., and Rine, J. (1988). Roles of two DNA-binding factors in replication, segregation and transcriptional repression mediated by a yeast silencer. *EMBO J.*, **7**, 2241–53.

Kipling, D., Ackford, H. E., Taylor, B. A., and Cooke, H. J. (1991). Mouse minor satellite DNA genetically maps to the centromere and is physically linked to the proximal telomere. *Genomics*, **11**, 235–41.

Kitsberg, D., Selig, S., Keshet, I., and Cedar, H. (1993). Replication structure of the human β-globin gene domain. *Nature*, **366**, 588–90.

Klar, A. J. S., Strathern, J. N., and Abraham, J. A. (1984). Involvement of double-strand chromosomal breaks for mating-type switching in *Saccharomyces cerevisiae*. *Cold Spring Harbor Symp. Quant. Biol.*, **49**, 77–88.

Kyrion, G., Boakye, K. A., and Lustig, A. J. (1992). C-terminal truncation of RAP1 results in the deregulation of telomere size, stability, and function in *Saccharomyces cerevisiae*. *Mol. Cell. Biol.*, **12**, 5159–73.

Kyrion, G., Liu, K., Liu, C., and Lustig, A. J. (1993). RAP1 and telomere structure regulate telomere position effects in *Saccharomyces cerevisiae*. *Genes Dev.*, **7**, 1146–59.

Lee, D. Y., Hayes, J. J., Pruss, D., and Wolffe, A. P. (1993). A positive role for histone acetylation in transcription factor access to nucleosomal DNA. *Cell*, **72**, 73–84.

Levis, R. W. (1989). Viable deletions of a telomere from a Drosophila chromosome. *Cell*, **58**, 791–801.

Longtine, M. S., Enomoto, S., Finstad, S. L., and Berman, J. (1992). Yeast telomere repeat sequence (TRS) improves circular plasmid segregation, and TRS plasmid segregation involves the *RAP1* gene product. *Mol. Cell. Biol.*, **12**, 1997–2009.

Longtine, M. S., Enomoto, S., Finstad, S. L., and Berman, J. (1993). Telomere-mediated plasmid segregation in *Saccharomyces cerevisiae* involves gene products required for transcriptional repression at silencers and telomeres. *Genetics*, **133**, 171–82.

McCarroll, R. M. and Fangman, W. L. (1988). Time of replication of yeast centromeres and telomeres. *Cell*, **54**, 505–13.

Mahoney, D. J., Marquardt, R., Shei, G.-J., Rose, A. B., and Broach, J. R. (1991). Mutations in the *HML* E silencer of *Saccharomyces cerevisiae* yield metastable inheritance of transcriptional repression. *Genes Dev.*, **5**, 605–15.

Micklem, G., Rowley, A., Harwood, J., Nasmyth, K., and Diffley, J. F. X. (1993). Yeast origin recognition complex is involved in DNA replication and transcriptional silencing. *Nature*, **366**, 87–9.

Miller, A. M. and Nasmyth, K. A. (1984). Role of DNA replication in the repression of silent mating type loci in yeast. *Nature*, **312**, 247–51.

Nasmyth, K. A. (1982). The regulation of yeast mating-type chromatin structure by *SIR*: an action at a distance affecting both transcription and transposition. *Cell*, **30**, 567–78.

Newlon, C. S., Collins, I., Dershowitz, A., Deshpande, A. M., Greenfeder, S. A., Ong, L. Y., and Theis, J. F. (1993). Analysis of replication origin function on chromosome III of *Saccharomyces cerevisiae*. *Cold Spring Harbor Symp. Quant. Biol.*, **58**, 415–23.

O'Keefe, R. T., Henderson, S. C. and Spector, D. L. (1992). Dynamic organization of DNA replication in mammalian cell nuclei: spatially and temporally defined replication of chromosome-specific α-satellite DNA sequences. *J. Cell Biol.*, **116**, 1095–110.

Palladino, F., Laroche, T., Gilson, E., Axelrod, A., Pillus, L., and Gasser, S.M. (1993). SIR3 and SIR4 proteins are required for the positioning and integrity of yeast telomeres. *Cell*, **75**, 543–55.

Park, E.-C. and Szostak, J. W. (1990). Point mutations in the yeast histone H4 gene prevent silencing of the silent mating type locus *HML*. *Mol. Cell. Biol.*, **10**, 4932–4.

Park, E.-C. and Szostak, J. W. (1992). ARD1 and NAT1 proteins form a complex that has N-terminal acetyltransferase activity. *EMBO J.*, **11**, 2087–93.

Pillus, L. and Rine, J. (1989). Epigenetic inheritance of transcriptional states in S. cerevisiae. *Cell*, **59**, 637–47.

Renauld, H., Aparicio, O. M., Zierath, P. D., Billington, B. L., Chhablani, S. K., and Gottschling, D. E. (1993). Silent domains are assembled continuously from the telomere and are defined by promoter distance and strength, and by *SIR3* dosage. *Genes Dev.*, **7**, 1133–45.

Rivier, D. H. and Rine, J. (1992). An origin of DNA replication and a transcriptional silencer require a common element. *Science*, **256**, 659–63.

Singh, J. and Klar, A. J .S. (1992). Active genes in budding yeast display enhanced in vivo accessibility to foreign DNA methylases: a novel in vivo probe for chromatin structure of yeast. *Genes Dev.*, **6**, 186–96.

Strathern, J. N., Klar, A. J. S., Hicks, J. B., Abraham, J. A., Ivy, J. M., Nasmyth, K. A., and McGill, C. (1982). Homothallic switching of yeast mating type cassettes is initiated by a double-stranded cut in the *MAT* locus. *Cell*, **31**, 183–92.

Sussel, L. and Shore, D. (1991). Separation of transcriptional activation and silencing functions of the *RAP1*-encoded repressor/activator protein 1: isolation of viable mutants affecting both silencing and telomere length. *Proc. Natl. Acad. Sci. USA*, **88**, 7749–53.

Sussel, L., Vannier, D., and Shore, D. (1993). Epigenetic switching of transcriptional states: *cis*-and *trans*-acting factors affecting establishment of silencing at the *HMR* locus in *Saccharomyces cerevisiae*. *Mol. Cell. Biol.*, **13**, 3919–28.

Ten Hagen, K. G., Gilbert, D. M., Willard, H. F., and Cohen, S. N. (1990). Replication timing of DNA sequences associated with human centromeres and telomeres. *Mol. Cell. Biol.*, **10**, 6348–55.

Venter, U. and Hörz, W. (1989). The acid phosphatase genes PHO10 and PHO11 in *S.cerevisiae* are located at the telomeres of chromosome VIII and I. *Nucleic Acids Res.*, **17**, 1353–69.

Weiffenbach, B. and Haber, J. E. (1981). Homothallic mating type switching generates lethal chromosome breaks in *rad52* strains of *Saccharomyces cerevisiae*. *Mol. Cell. Biol.*, **1**, 522–34.

Wellinger, R. J., Wolf, A. J., and Zakian, V. A. (1993). Origin activation and formation of single-strand TG_{1-3} tails occur sequentially in late S phase on a yeast linear plasmid. *Mol. Cell. Biol.*, **13**, 4057–65.

Wines, D. R. and Henikoff, S. (1992). Somatic instability of a Drosophila chromosome. *Genetics*, **131**, 683–91.

Wright, J. H., Gottschling, D. E., and Zakian, V. A. (1992). *Saccharomyces* telomeres assume a non-nucleosomal chromatin structure. *Genes Dev.*, **6**, 197–210.

Wu, L.-C. C., Fisher, P. A., and Broach, J. R. (1987). A yeast plasmid partitioning protein is a karyoskeletal component. *J. Biol. Chem.*, **262**, 883–91.

9

Structure and maintenance of *Drosophila* telomeres

Drosophila melanogaster has a telomeric sequence organization and mechanism of sequence addition dramatically different from other species which have been analysed. From what has been reviewed up to this point one might think that all eukaryotic chromosomal telomeres consist of a tandem repeat array based on a short sequence, probably synthesized by telomerase. *Candida albicans* is unusual, as its terminal repeats are 23 bp in length, but it is still possible that they are synthesized by telomerase. However, *Drosophila* provides the most clear exception to this type of telomere structure and synthesis (reviewed by Biessmann and Mason 1992; Mason and Biessmann 1993).

Many insects do have a short terminal repeat sequence, which is $(TTAGG)_n$ (Okazaki *et al.* 1993). However, a simple repeat sequence has yet to be found at the end of *Drosophila* chromosomes despite numerous attempts. Repeats from other species have been used in attempts to detect BAL31-sensitive signals by cross-hybridization, plasmid end-libraries have been constructed, and telomere cloning in yeast has been attempted. None of these approaches have identified a candidate short terminal repeat sequence. Instead, there is a large body of evidence to suggest that *Drosophila* chromosomes terminate in arrays of retroposon-like sequences called HeT-A elements. These appear to be mobile elements with a strong preference for transposition to the ends of DNA molecules. Frequent transposition events to the ends of *Drosophila* chromosomes would add sufficient sequence to overcome that lost because of the end-replication problem. Another interesting feature is that natural chromosome ends in *Drosophila* do not appear to be prone to end-to-end fusion or other recombination events, irrespective of what sequence they terminate in. This is in direct contrast to the situation in other species which have been discussed, where specific terminal repeat sequences are implicated in protecting the natural chromosomal termini from such deleterious processes. How *Drosophila* protects its ends in the absence of a specific sequence element remains an unanswered question.

Terminal deletions can be obtained in *Drosophila*

The telomere was first defined by Muller (1938) on the basis of the very low frequency at which terminal deletions and inversions of *Drosophila* chromosomes are found following mutagenesis (Chapter 2). This contrasts with the relative ease with which internal deletions or inversions can be obtained. The paucity of deletions that included removal of the end of the chromosome suggested the presence of a special terminal structure essential for chromosome function—the telomere. Studies of breakage–fusion–bridge cycles in maize suggested that one essential role was to prevent natural termini from fusing with other regions of the genome. Almost all subsequent studies have supported the hypothesis that a free end is prone to recombination with the rest of the genome, and that linear molecules can only be maintained even in the short term if they are capped by terminal repeat sequences.

Terminal deletions of chromosomes have been obtained in a number of species (Chapter 6) but in all cases analysed they have been 'healed' by the addition of telomeric sequences, thus acquiring telomere function. In contrast, it now appears that *Drosophila* is one species where terminal deletions can be obtained which are *not* healed by the addition of specific telomeric sequences but by some other process, perhaps binding of a protein complex (see below). Two mechanisms for efficiently producing such deletions have been described.

One way is by P element induced chromosome breakage. Levis (1989) has recovered a number of terminal deletions by introducing a source of P transposase, causing the destabilization of a P element transposon located very near the end of a chromosome. The P element contained a *white* gene, the expression of which was subject to position effect variegation caused by the heterochromatin adjacent to the insertion site. The result of P transposase expression was a double-strand break between the P element and the terminus of the chromosome. In flies which had lost the distal chromosomal material the position effect was relieved, resulting in stable *white* expression in the eye. These terminal deletions have been analysed using probes for P element sequences, which are now at the terminus of the chromosome. The genomic restriction maps indicate that these deletions are not accompanied by the acquisition of any new sequences at the breakpoints; the resolution of this analysis is such that if the chromosome has acquired a new terminal sequence it would need to be less than 100 bp in size. Furthermore, the terminal restriction fragments, which are composed of P element sequences, shorten at the rate of around 50–100 bp per fly generation (Levis 1989). This also argues against a short terminal repeat sequence having been added to the broken chromosome because it would require this terminal repeat sequence to be maintained while at the same time there being progressive loss of the subterminal P element sequences.

Terminal deletions can also be found in the offspring of X-irradiated females homozygous for the *mutator-2* (*mu-2*) mutation (Mason *et al.* 1984). This mutation, on the left arm of the third chromosome, may either prevent efficient repair of double-strand breaks, or may overcome the cell cycle arrest that is a likely response to X-ray induced damage, thus allowing cells to divide which have not repaired the break. Whatever the mechanism, once formed the double-strand break becomes stable in that such terminally deleted chromosomes can be maintained in wild-type flies without activating cell cycle arrest or being repaired. The *mu-2*-facilitated terminal deletions which have been studied in the most detail are those that remove the tip of the X chromosome (Biessmann and Mason 1988; Biessmann *et al.* 1990*b*). These were found by screening for loss of expression of the *yellow* gene, which is close to the end of the X chromosome. Death because of loss of essential genes distal to *yellow* in the terminal deficiencies was avoided by using a strain carrying a translocation chromosome carrying the tip of the X chromosome on the Y. X chromosome terminal deletions can be obtained at high (0.2–0.3%) frequency by X-irradiating *mu-2* homozygous females. A number of these deletions have been obtained and analysed in detail using probes for the *yellow* locus. Restriction mapping and cloning terminal restriction fragments suggests that these deletions terminate in *yellow* sequences; there is no evidence that they have acquired any new sequence at the terminus. They show sequence loss at the distal ends at a rate of about 75 bp per fly generation.

The rate of sequence loss is consistent with incomplete replication

One can calculate from what is known about DNA replication the minimum rate of sequence loss that the end-replication problem would cause. For example, consider the most conservative situation where the RNA primer starts at the first nucleotide of the chromosome. One of the four daughter strands would lose sequence equivalent to the length of this primer (see Fig. 1.1). Therefore the minimum loss rate at any terminus is one-quarter of the length of the primer per cell division. In *Drosophila* the size of the RNA primer has been measured to be 8 nt (Kitani *et al.* 1984) so the minimum loss rate is 2 bp per cell division. It has been estimated that there is a minimum of 17–19 cell divisions in the germ-line lineages of both males and females, thus the minimum loss rate is 34–38 bp per fly generation. The observed rate of sequence loss for chromosomes with terminal deletions is therefore at least as large as the minimum predicted by the end-replication problem. This is important because if the observed rate of loss were lower this would indicate a process actively adding sequences to the termini. However, the measured rate is consistent with a progressive loss of sequence because of incomplete replication. A similar rate of loss is also observed for natural chromosome ends, and long-term length stability is achieved by the occasional additions of a large amount of sequence to the ends in the form of HeT-A transposons (see below).

HeT-A sequences can transfer to receding ends

Various different fly strains with receding chromosome ends have been maintained for a number of years and continue to show progressive loss of terminal sequences. However, occasionally flies arise which cease to show loss of *yellow* sequences. In the eight examples which have been analysed in detail the terminal deletions have acquired new genomic sequences; *yellow* sequences are no longer lost because there is now additional sequence distal to the gene. These sequences are similar to those which occur at natural *Drosophila* telomeres and are members of a repetitive family called HeT-A (Biessmann *et al.* 1990*a*). HeT-A is a member of a family of repeated sequences called HeT DNA, originally defined using a clone called λT-A, and are found in the telomeric and pericentromeric heterochromatin of *Drosophila* (Young *et al.* 1983). By *in situ* hybridization HeT sequences have been found at all telomeres (Rubin 1978; Young *et al.* 1983; Traverse and Pardue 1989; Valgeirsdottir *et al.* 1990; Danilevskaya *et al.* 1991, 1992) as well as new ends caused by spontaneous opening of a ring chromosome (Traverse and Pardue 1988).

HeT-A is a set of transposable elements, and it is these elements that have transferred themselves to the receding terminal deletions (Biessmann *et al.* 1992*a,b*). The ends of broken X chromosomes with HeT-A additions recede at the same rate as the chromosomes without HeT-A terminal sequences; the addition of HeT-A sequences does not prevent terminal sequence loss. The addition of HeT-A sequences to any particular receding chromosome occurs at a frequency of approximately 1% per generation (Biessmann *et al.* 1992*b*) with on average 6 kb of sequence being added. This rate of HeT-A addition is therefore equivalent to about 60 bp of sequence per fly generation and therefore would appear to be sufficient to counteract the observed rate of terminal sequence loss (70–75 bp per generation) caused by incomplete replication.

HeT-A is a mobile element by definition of it having been transferred onto the ends of broken chromosomes. Sequence comparison of the target sites suggests that this movement does not occur by homologous recombination, nor does there appear to be any primary sequence specificity for the target site. One end of each HeT-A element is a 3' poly(A) tail of variable length, 53 bp downstream of a polyadenylation signal. It has therefore been suggested that HeT-A elements are transposons whose movement involves a poly(A)$^{+}$ RNA intermediate. As transposition would therefore require a step mediated by reverse transcriptase this would classify HeT-A as a retroposon, similar to mobile elements such as the *Drosophila* jockey element or mammalian LINEs. Most genomic copies of HeT-A appear truncated at the 5' end but the full-length copy is estimated to be approximately 6 kb in size (Valgeirsdottir *et al.* 1990; Biessmann *et al.* 1992*a,b*, 1994; Danilevskaya *et al.* 1992, 1993). Such long clones contain an open reading frame which encodes a 918 amino acid *gag*-like protein with three zinc finger motifs and a high similarity to the *gag*-like protein encoded by the *Drosophila* retroposon jockey, consistent with HeT-A also being a retroposon. Retroposons typically contain two open reading frames, one of which contains a *gag*-like protein and the other a reverse transcriptase. In this respect HeT-A is atypical as even full-length clones do not appear to contain an open reading frame for a reverse transcriptase. One speculation is that this activity is encoded by a non-HeT-A sequence and is provided *in trans*; one candidate for this source of reverse transcriptase is the TART element (see below). The presumed reliance on reverse transcriptase encoded elsewhere in the genome distinguishes HeT-A elements from 'selfish' LINE-like elements (Biessmann *et al.* 1994).

HeT-A elements can move to locations where there is no pre-existing element. Do HeT-A elements move to telomeres because they are mobile elements with a preference for telomeric heterochromatin? There are precedents for transposons with a preference for where they move to (Plasterk 1993). HeT-A elements are usually found only in telomeric and pericentromeric heterochromatin, but they can be found in euchromatic regions provided they are telomeric, such as the *yellow* gene if a terminal deletion occurs within it, or the new ends formed following opening of a ring chromosome (Traverse and Pardue 1988; Biessmann *et al.* 1990*a*, 1992*a,b*).

There is no evidence from the analysis of HeT-A insertions sites that target sites are chosen on the basis of primary sequence. Instead the target specificity may be for DNA topology, in the form of a double-strand break. All the analysed transposition events of HeT-A elements to terminally deleted chromosomes have the poly(A) tail at the boundary between the HeT-A element and the broken chromosome end (Biessmann *et al.* 1992*b*). A mechanism has been proposed for the transposition of retroposons of the class to which HeT-A may belong. It is hypothesized that the 3' end of the poly(A)$^+$ RNA intermediate forms a complex with reverse transcriptase and the *gag*-like protein. This complex attaches the 3' poly(A) tail to a nick in the genomic DNA and reverse transcription is primed using the 3'-OH of the nicked DNA. By analogy it has been suggested (Mason and Biessmann 1993; Levis *et al.* 1993) that the poly(A) tail of the HeT-A intermediate is attached to the terminal double-strand break of the chromosome and this is used to initiate reverse transcription. This provides an explanation for the highly polar organization of HeT-A elements. One further prediction is that naturally occurring subtelomeric HeT-A elements would be found with the poly(A) tails located proximally; this appears to be the case for those elements which have been analysed (Biessmann *et al.* 1992*a*, 1993, 1994; Karpen and Spradling 1992).

The evidence suggests that HeT-A is a retroposon which transposes to free DNA ends. From the rates of transposition (*c.* 1% per generation) and the average 6 kb that is transferred, the observed loss of sequence would predict a dynamic turnover of HeT-A elements and the formation of a tandem repeat array of HeT-A elements which are irregular in size, with varying truncations at the 5′ end, separated by poly(A) tails and all oriented in the same polar fashion (Fig. 9.1). Analysis of a recently formed array added to a receding chromosome (Biessmann *et al.* 1994) reveals just such an arrangement.

A natural chromosome end has recently been cloned by Levis *et al.* (1993). In addition to HeT-A sequences they also found another retroposon-like element they call TART. The two elements are very different in sequence, despite some similarities in the open reading frame for the *gag*-like protein, eliminating the possibility that one element originated from the other. A major difference between HeT-A and TART is that the latter contains an open reading frame for a reverse transcriptase. By *in situ* hybridization TART elements are not found at all *Drosophila* telomeres, and cloning the entire subtelomeric region of 2L reveals only HeT-A elements (H. Biessmann, personal communication). In all the X chromosome healing events studied so far only HeT-A elements have been found; as yet there is no evidence that TART is an active transposon and can jump. However, only HeT-A elements may have been seen to jump so far because they are more numerous in the genome than TART, and it cannot be discounted that there exist additional LINE-like elements in *Drosophila*, such as TART, which behave in a similar fashion to HeT-A.

Consequences of the retroposon model of telomere maintenance

The retroposon model suggests that the rate of terminal sequence loss caused by the end-replication problem is balanced by an equal rate of sequence addition in the form of duplicative transpositions of HeT-A mobile elements to terminal double-strand breaks. If the addition of HeT-A elements to ends is stochastic one might expect the occasional loss of all HeT-A sequences at a natural telomere and subsequent loss of some distal euchromatic sequence. Evidence in support of this model is that small terminal deletions do occur naturally. The *lethal(2)giant larvae* (*l(2)gl*) gene is the most distal on chromosome 2L and 1–2% of flies in the wild carry a heterozygous *l(2)gl* mutation (discussed by Biessmann and Mason 1992). Molecular analysis reveals that the majority of these are truncations that remove the 3′ region of the gene, which is transcribed towards the telomere (Mechler *et al.* 1985). This is consistent with occasional, stochastic terminal deletions. Larger deletions may also occur but they will be eliminated

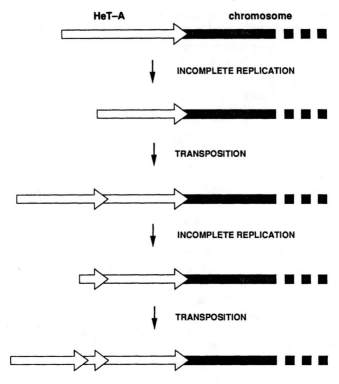

Fig. 9.1 Model for telomere maintenance by HeT-A transposition in *Drosophila*. HeT-A elements are indicated by arrows, with the arrowhead denoting the poly(A) tail pointing away from the end of the chromosome. Incomplete replication leads to shortening of the terminal HeT-A element. Transposition of new HeT-A elements occurs to the chromosome terminus, possibly in a stochastic manner. Telomere length maintenance is a balance between continual loss and occasional addition of a large amount of HeT-A sequence. The final result is a terminal array of HeT-A elements, all with the same orientation and showing varying degrees of 5′ truncation. Reprinted with permission from Mason and Biessmann (1993), © 1993 Springer-Verlag, Berlin.

from the population if they remove genes for which hemizygosity is deleterious.

How does *Drosophila* survive?

There is considerable evidence that free ends do pose a problem for *Drosophila* and that it is not tolerant of double-strand breaks. The high rates of end-to-end fusions and the genomic rearrangements resulting from the subsequent formation of dicentric chromosomes would not be compatible with survival of the organism. How then does *Drosophila* protect the ends of its chromosomes in the absence of a specific DNA sequence? One potential explanation might be that *Drosophila* telomeres are heterochromatic, and a free DNA end in such a region is assembled into a different chromatin structure and is therefore inaccessible to recombination proteins. However, this hypothesis seems unlikely as the viable terminal deletions in the *yellow* locus produce chromosome ends which are visibly euchromatic. Another possibility is that end-to-end fusions could occur but these are then subject to an efficient processing reaction to resolve them. However, as *Drosophila* chromosomes can terminate in a variety of different sequences, including non-HeT-A sequences, it is not clear just what molecular structure could possibly be recognized by the resolution reaction.

One of the most puzzling aspects of these terminal deletions is why they can be obtained in the first place, and especially how they can be maintained in wild-type flies without repair or activation of cell cycle checkpoints. It has been suggested that *mu-2* prevents the repair of the initial X-ray-induced double-strand break, either directly or by allowing cells with an unrepaired break to continue cell division. However, once formed the break can be passed through the female germ-line without activating arrest or repair; why the different cellular response to an 'old' as opposed to a 'new' double-strand break? It has been suggested (Biessmann and Mason 1988) that this might be explained by a protein which binds in a sequence-independent fashion to double-strand breaks. This would bind the termini of the chromosomes and protect them from fusion, and would be postulated to recognize double-strand breaks rather than a specific telomeric sequence. A newly-formed double-strand break, initially not complexed with this protein, would be recognized as a 'new' break by the cell and be a target for repair processes or recombination. However, if the repair processes can be delayed sufficiently (e.g. by *mu-2*) then a protein complex may form on this new end, which will now be recognized as a 'telomere' and not be subject to subsequent repairs. The delay in assembling this DNA-protein complex provides one explanation why only a newly-formed X-ray induced double-strand break triggers repair or cell cycle arrest.

What then is a *Drosophila* telomere? The work of Muller clearly indicated that the end of the chromosome has a specialized, functionally important structure. The stability of terminal deletions within the *yellow* gene suggests that the structure required for stability does not reflect a specific DNA sequence; however, a telomere is defined functionally and not necessarily by the presence of a certain class of DNA sequences. Functionally a *Drosophila* telomere is different from a simple double-strand break, possibly because it is complexed with specific proteins which protect that end from recombination and degradation. The processes of protecting the end of the chromosome from fusion and addition of sequences to compensate for incomplete replication appear to be separate in *Drosophila*. The capping function does not require a specific DNA sequence, whereas the elongation mechanism uses HeT-A (and possibly TART) elements. A *Drosophila* telomere is a functional entity that cannot be defined purely on the basis of DNA sequence.

Similarities to and differences from telomere maintenance in yeast *est1* mutants

It is interesting to compare a mechanism of telomere maintenance which occurs in *S. cerevisiae est1* mutants with the situation postulated to occur in *Drosophila*. The *est1* mutation identifies a gene, described in detail in Chapter 5, which may encode a protein component of the as yet unidentified yeast telomerase enzyme. *Est1* mutants show progressive loss of the terminal TG_{1-3} repeats and eventually senesce. However, it is common for cultures in the late stages of senescence (about 100 generations) to be overgrown by rapidly growing cells. These survivors also occur in an *est1*Δ deletion mutant, indicating they are not caused by reversion at the *est1* locus. These survivors still have terminal TG_{1-3} tracts and have therefore activated some mechanism to prevent further sequence loss.

The mechanism of terminal repeat maintenance in these survivors has been analysed in detail by Lundblad and Blackburn (1993). These survivors have undergone amplification of the subtelomeric Y' elements. This amplification is probably caused by recombination because these survivors do not occur in a *rad52* strain, a mutation which abolishes most types of mitotic recombination. Recombination between Y' elements at different locations has been observed and can lead to expansion of single copies into tandem repeat arrays (Louis and Haber 1990). This expansion could occur by unequal sister chromatid exchange or gene conversion (Fig. 9.2(a)). Y' elements have also been detected as extrachromosomal circles, which could be intermediates in their movement (Fig. 9.2(b)). These *RAD52*-dependent recombinational events probably underlie the amplification of Y' elements which occurs in *est1* survivors.

Once the number of Y' elements has been amplified in the cell, a new mechanism of telomere maintenance is suggested to occur. Y' elements are flanked on both sides by tracts of TG_{1-3} and have been observed to transfer to a chromosome end which does not possess a Y' element (Louis and Haber 1990). These movements could occur by either of the two mechanisms shown in Fig. 9.2, and could also lead to a net increase in the amount of terminal TG_{1-3} sequence. Normally these processes would be too rare to compensate for terminal sequence loss; Y' elements recombine with each other at a rate of around 10^{-6} per cell per generation (Louis and Haber 1990). However, by amplifying the total number of Y' elements in the genome the frequency of events such as those shown in Fig. 9.2 may be increased.

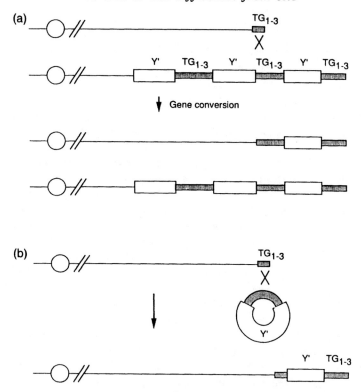

Fig. 9.2 Telomere sequence addition in *est1⁻* survivors. (a) Gene conversion between two telomeres initiated by a terminal TG$_{1-3}$ tract invading an internal TG$_{1-3}$ tract at another telomere. (b) Circular Y' elements can be excised from arrays of Y' elements and then recombine with an under-replicated telomere. Reprinted with permission from *Cell* (Lundblad and Blackburn 1993), © 1993 Cell Press.

Cells which contain amplified Y' elements are selected for because they enable the cell to survive the *est1* mutation. An *est1* survivor transformed with a plasmid carrying the *EST1* gene, thus making the cell wild-type, rapidly loses its amplified Y' elements. It therefore seems that under normal growth conditions amplified Y' elements, which can constitute a significant fraction of the total genome in the *est1Δ* survivors, confer a selective disadvantage. It is only their ability to rescue a cell from the lethality of an *est1* mutation that enables their maintenance in a survivor.

Although there are parallels between this mechanism of telomere sequence addition and that which occurs in *Drosophila*, there are two crucial differences. Firstly, although the sequence of Y' suggests that it might have had its origin as a mobile element, there is no evidence to suggest that it can currently move by any mechanism which does not involve a recombination step. Secondly, the consequence of Y' element movement is that it also transfers TG_{1-3} terminal repeats, thus arresting the progressive loss of these sequences in the *est1* mutant. In contrast there is no evidence for a short terminal repeat sequence in *Drosophila*, and not only does HeT-A appear to move by transposition but its addition alone may be sufficient for telomere maintenance.

How many ways are there to replicate a telomere?

There is little direct evidence that recombination has a role in the day-to-day maintenance of terminal repeat sequences in yeast or any other species (Chapter 6). However, the telomere maintenance pathway that is revealed when the normal pathway is abolished by the *est1* mutation indicates that recombination-based mechanisms can, in principle, provide sufficient sequence addition to chromosomal termini to overcome sequence loss because of incomplete replication. The models suggested for this pathway (Fig. 9.2) result in chromosomes maintaining their terminal repeat sequences.

Drosophila, however, provides an example to illustrate that a specific terminal sequence is not necessary to stabilize a chromosome against fusion and degradation. Therefore, if a species can similarly protect natural chromosome ends in the absence of a specific DNA sequence then recombination-based pathways could readily be invoked to produce terminal sequence addition to overcome that lost by incomplete replication. For example, it might be possible to utilize similar mechanisms to those that are responsible for the expansion of the tandem repeat arrays which form many heterochromatin domains. The mechanism by which such arrays amplify remains controversial, with unequal crossover and extrachromosomal rolling-circle mechanisms being two of the many suggestions. Whatever genetic mechanism is capable of amplifying a large tandem repeat array, such as the α-satellite found at human centromeres, would also in principle be capable of adding sequence to a terminal heterochromatin block.

Drosophila appears to amplify its HeT-A elements by duplicative transposition rather than a recombination-based mechanism. How common is such a transposase-based mechanism? $(TTAGG)_n$ is found in many but not all insects (Chapter 3) and it is possible that other species which do not appear to have a simple terminal repeat sequence also use a transposon-based telomere maintenance pathway. However, such terminally located transposons have not yet been reported in other species. What is known is that many plant and animal species have large blocks of terminal heterochromatin composed of tandem repeat arrays. It will be of interest to determine if any take advantage of these to provide a mechanism of terminal sequence addition as opposed to a telomerase-based mechanism to add short terminal repeat sequences.

References

Reviews

Biessmann, H. and Mason, J. M. (1992). Genetics and molecular biology of telomeres. *Adv. Genet.*, **30**, 185–249.

Mason, J. M. and Biessmann, H. (1993). Transposition as a mechanism for maintaining telomere length in *Drosophila. Chromosome segregation and aneuploidy.* NATO ASI, Series H, pp. 143–9. Springer-Verlag, Berlin.

Primary papers

Biessmann, H. and Mason, J. M. (1988). Progressive loss of DNA sequences from terminal chromosome deficiencies in *Drosophila melanogaster. EMBO J.*, **7**, 1081–6.

Biessmann, H., Mason, J. M., Ferry, K., d'Hulst, M., Valgeirsdottir, K., Traverse, K. L., and Pardue, M.-L. (1990a). Addition of telomere-associated HeT DNA sequences 'heals' broken chromosomes ends in Drosophila. *Cell*, **61**, 663–73.

Biessmann, H., Carter, S. B., and Mason, J. M. (1990b). Chromosome ends in *Drosophila* without telomeric DNA sequences. *Proc. Natl. Acad. Sci. USA*, **87**, 1758–61.

Biessmann, H., Valgeirsdottir, K., Lofsky, A., Chin, C., Ginther, B., Levis, R. W., and Pardue, M.-L. (1992a). HeT-A, a transposable element specifically involved in 'healing' broken chromosome ends in *Drosophila melanogaster. Mol. Cell. Biol.*, **12**, 3910–18.

Biessmann, H., Champion, L. E., O'Hair, M., Ikenaga, K., Kasravi, B., and Mason, J. M. (1992b). Frequent transpositions of *Drosophila melanogaster* HeT-A transposable elements to receding chromosome ends. *EMBO J.*, **11**, 4459–69.

Biessmann, H., Kasravi, B., Jakes, K., Bui, T., Ikenaga, K. and Mason, J. M. (1993). The genomic organization of HeT-A retroposons in *Drosophila melanogaster. Chromosoma*, **102**, 297–305.

Biessmann, H., Kasravi, B., Bui, T., Fujiwara, G., Champion, L.E., and Mason, J. M. (1994). Comparison of two active HeT-A retroposons of *Drosophila melanogaster. Chromosoma*, **103**, 90–8.

Danilevskaya, O. N., Kurenova, E. V., Pavlova, M. N., Bebehov, D. V., Link, A. J., Koga, A., *et al.* (1991). He-T family DNA sequences in the Y chromosome of *Drosophila melanogaster* share homology with the X-linked stellate genes. *Chromosoma*, **100**, 118–24.

Danilevskaya, O. N., Petrov, D. A., Pavlova, M. N., Koga, A., Kurenova, E. V., and Hartl, D. L. (1992). A repetitive DNA element, associated with telomeric sequences in *Drosophila melanogaster*, contains open reading frames. *Chromosoma*, **102**, 32–40.

Danilevskaya, O., Lofsky, A., Kurenova, E. V., and Pardue, M.-L. (1993). The Y-chromosome of *Drosophila melanogaster* contains a distinctive subclass of HeT-A-related repeats. *Genetics*, **134**, 531–43.

Karpen, G. H. and Spradling, A. C. (1992). Analysis of subtelomeric heterochromatin in the Drosophila minichromosome *Dp1187* by single P element insertional mutagenesis. *Genetics*, **132**, 737–53.

Kitani, T., Yoda, K.-Y., and Okazaki, T. (1984). Discontinuous DNA replication of *Drosophila melanogaster* is primed by octaribonucleotide primer. *Mol. Cell. Biol.*, **4**, 1591-6.

Levis, R. W. (1989). Viable deletions of a telomere from a Drosophila chromosome. *Cell*, **58**, 791-801.

Levis, R. W., Ganesan, R., Houtchens, K., Tolar, L. A., and Sheen, F.-m. (1993). Transposons in place of telomeric repeats at a Drosophila telomere. *Cell*, **75**, 1083-93.

Louis, E. J. and Haber, J. E. (1990). Mitotic recombination among subtelomeric Y′ repeats in *Saccharomyces cerevisiae*. *Genetics*, **124**, 547-59.

Lundblad, V. and Blackburn, E. H. (1993). An alternative pathway for yeast telomere maintenance rescues *est1*⁻ senescence. *Cell*, **73**, 347-60.

Mason, J. M., Strobel, E., and Green, M. M. (1984). *mu-2*: mutator gene in *Drosophila* that potentiates the induction of terminal deficiencies. *Proc. Natl. Acad. Sci. USA*, **81**, 6090-4.

Mechler, B. M., McGinnis, W., and Gehring, W. J. (1985). Molecular cloning of *lethal(2)giant larvae*, a recessive oncogene of *Drosophila melanogaster*. *EMBO J.*, **4**, 1551-7.

Muller, H. J. (1938). The remaking of chromosomes. *Collect. Net*, **8**, 182-95.

Okazaki, S., Tsuchida, K., Maekawa, H., Ishikawa, H., and Fujiwara, H. (1993). Identification of a pentanucleotide telomeric sequence, $(TTAGG)_n$, in the silkworm *Bombyx mori* and in other insects. *Mol. Cell. Biol.*, **13**, 1424-32.

Plasterk, R. H. A. (1993). Molecular mechanisms of transposition and its control. *Cell*, **74**, 781-6.

Rubin, G. M. (1978). Isolation of a telomeric DNA sequence from *Drosophila melanogaster*. *Cold Spring Harbor Symp. Quant. Biol.*, **42**, 1041-6.

Traverse, K. L. and Pardue, M. L. (1988). A spontaneously opened ring chromosome of *Drosophila melanogaster* has acquired He-T DNA sequences at both new telomeres. *Proc. Natl. Acad. Sci. USA*, **85**, 8116-20.

Traverse, K. L. and Pardue, M. L. (1989). Studies of He-T DNA sequences in the pericentric regions of *Drosophila* chromosomes. *Chromosoma*, **97**, 261-71.

Valgeirsdottir, K., Traverse, K. L., and Pardue, M.-L. (1990). HeT DNA: a family of mosaic repeated sequences specific for heterochromatin in *Drosophila melanogaster*. *Proc. Natl. Acad. Sci. USA*, **87**, 7998-8002.

Young, B. S., Pession, A., Traverse, K. L., French, C., and Pardue, M. L. (1983). Telomere regions in Drosophila share complex DNA sequences with pericentric heterochromatin. *Cell*, **34**, 85-94.

10

Telomeres and mammalian genome analysis

This chapter reviews the cloning and sequence organization of human telomeres, their integration with the genetic map, and how their structure compares with those in the mouse. Evidence for true terminal deletions, clustering of VNTR (variable number of tandem repeats) loci in proterminal regions, terminal map expansions, and telomere–telomere exchanges in humans will be reviewed, as will the ability to manipulate chromosome structure experimentally by telomere-associated chromosome fragmentation. Some internal $(TTAGGG)_n$ arrays may be the remnants of ancestral telomere fusion events, which will be illustrated by the evolution of human chromosome 2. Telomeric sequences can be used to investigate a number of aspects of chromosome structure and behaviour, such as fragile sites, and in particular can reveal cryptic translocations in patients with apparently normal karyotypes. Finally, a potential link between the unusual genetic behaviour of $(TTAGGG)_n$ repeat arrays, as exemplified by those of the mouse, and the instability of short repeated sequences which underlie some human genetic diseases will be discussed.

Terminal repeat array structure in humans

The presence of short, repeated sequences at ciliate telomeres led to a search for similar repeats at the ends of human chromosomes. The first success came when Allshire *et al.* (1988) were able to detect a BAL31-sensitive signal when probing Southern blots of digested human DNA with $(TTGGGG)_n$ under stringent conditions. However, subsequently other related sequences were also reported that would hybridize to human telomeres. For example, $(TTAGGG)_n$ produces a telomeric signal on metaphase chromosomes by *in situ* hybridization (Moyzis *et al.* 1988). Similarly, a probe corresponding to the *Arabidopsis* terminal repeats, $(TTTAGGG)_n$, also detects a BAL31-sensitive signal in humans (Richards and Ausubel 1988). Therefore it was not clear which sequence was the authentic terminal repeat, especially given the ability of these short G-rich sequences to cross-hybridize to each other (Chapter 3).

This was resolved by cloning human telomeres in yeast and *E. coli* and demonstrating that they contain $(TTAGGG)_n$. Repeats in the more internal region of some of the human terminal repeat tracts cloned in yeast are degenerate, with TTGGGG, TGAGGG, TAGGG, and TTAGGG the main variants (Brown *et al.* 1990). This confirms earlier hybridization data indicating the presence of TGAGGG and TTGGGG repeats subterminally (Allshire *et al.* 1989). (TTAGGG)$_n$ had been known for many years to be a satellite sequence in the kangaroo rat (Fry and Salser 1977) and guinea pig (Southern 1970). The (TTAGGG)$_n$ signal in humans is a smear on Southern blots, some 5–15 kb in size for DNA from lymphocytes and some 10 kb larger for sperm DNA (Cooke and Smith 1986; Allshire *et al.* 1988, 1989; Cross *et al.* 1989; de Lange *et al.* 1990; Hastie *et al.* 1990). In contrast, the yeast clones inevitably contain less than 1 kb of (TTAGGG)$_n$. As with cloning telomeres from other species (see Chapter 3) chromosome stability in yeast is provided by the addition of yeast TG$_{1-3}$ repeats onto the human (TTAGGG)$_n$ repeats. One speculation is that the shortness of the (TTAGGG)$_n$ arrays obtained reflects this addition of TG$_{1-3}$ repeats being inefficient, so that much of the (TTAGGG)$_n$ array is lost by, for example, exonuclease activity before healing occurs. Alternatively, large (TTAGGG)$_n$ arrays may be inherently unstable in yeast. Whatever the reason, the available sequence data may correspond to regions of the array that are over 5 kb from the end of the chromosome, and it therefore remained a formal possibility that another sequence is at the very end of the chromosome. This contrasts to the situation in *S. cerevisiae* (Chapter 3) where the terminal repeat arrays are so short that plasmid clones containing sequences almost at the very end of the chromosome can be obtained. However, the observation that human telomerase adds (TTAGGG)$_n$ repeats *in vitro* (Chapter 4) is very strong evidence that human chromosomes end in this sequence.

Closing the human genetic map with cloned human telomeres

One major goal of the human genome mapping project is to 'close' the physical and genetic maps of chromosomes by including their ends, and this will require the isolation of sequences close to the telomeres. (TTAGGG)$_n$ tracts can be cloned directly in plasmid vectors (Moyzis *et al.* 1988) together with short stretches of subtelomeric sequence (de Lange *et al.* 1990). Small stretches of (TTAGGG)$_n$-adjacent sequence can also be isolated by PCR-based approaches (Royle *et al.* 1992) although none of the clones identified in one such strategy detected BAL31-sensitive fragments (Weber *et al.* 1990). However, such clones are relatively small, which may present a problem if such sequences are to be integrated with existing cosmid and YAC contigs. Because of the way conventional YAC and cosmid libraries are constructed, only fragments flanked on both sides by restriction sites will be cloned, and thus terminal restriction fragments will not be present in such libraries. A telomere-derived clone may therefore need to include substantial amounts of the subtelomeric region if it is to overlap with conventional cosmid or YAC clones. Furthermore, the presence of significant amounts of subtelomeric repetitive sequences, some present at many telomeres, can make the isolation of chromosome-specific probes from the immediate subterminal region problematic. For example, the terminal 60 kb of human chromosome 4p contains sequences homologous to those at some other telomeres (Youngman *et al.* 1992). In addition, sequences present at other telomeres can also be found up to about 285 kb internal to the physical end of the short arm of chromosome 16 (Wilkie *et al.* 1991; Wilkie and Higgs 1992).

These problems have been circumvented by cloning much larger terminal fragments in yeast using 1/2-YAC telomere cloning approaches, as described in Chapter 3 (Brown 1989; Cheng *et al.* 1989; Cross *et al.* 1989; Riethman *et al.* 1989; Bates *et al.* 1990). Chromosome-specific probes have been isolated from many of these clones and in some cases have been used to integrate these clones with genetic and physical maps. These include probes for the telomeres of 4q (Weber *et al.* 1990) and 9q (Guerrini *et al.* 1990). The telomeres of 16p (Wilkie *et al.* 1991), 4p (Bates *et al.* 1990; Youngman *et al.* 1992), 7q (Riethman *et al.* 1989, 1993; Brown *et al.* 1990; Dietz-Band *et al.* 1990; Helms *et al.* 1992) and the pseudoautosomal telomere (Brown 1988; Brown *et al.* 1990; Henke *et al.* 1993) have been characterized in particular detail.

Human subtelomeric sequences

A few kilobases of subtelomeric sequence are available for a number of chromosomes, and these show a complex selection of repetitive sequences. They are predominantly short repeats, with examples that are variously 29, 37, 46, and 61 bp in length, often in the form of short tandem repeats with differing periodicities (Brown *et al.* 1990; Cross *et al.* 1990; de Lange *et al.* 1990; Weber *et al.* 1990, 1991*b*; Cheng *et al.* 1991). They are present in variable copy number and can be entirely absent from some telomeres. Some of these repeats are very $G + C$-rich, such as one 29 bp repeat that is over 85% $G + C$ and acts as a hypervariable minisatellite restricted to primate species (Cross *et al.* 1990). An additional family of subtelomeric repeats from many chromosome ends has been described (Rouyer *et al.* 1990) although their location is unclear; they do not detect BAL31-sensitive signals on Southern blots and would therefore appear to be at least 20 kb from the chromosome ends. None of the reported subtclomeric sequences are found at all the telomeres, which argues against a role in essential telomere function. Only $(TTAGGG)_n$ is found at all human telomeres; further evidence for this sequence being sufficient for essential telomere function comes both from two examples of α-thalassaemia patients with terminal deletions healed by addition of $(TTAGGG)_n$ and no other sequence (Chapter 6) and the ability of cloned $(TTAGGG)_n$ to heal chromosome breaks in cultured cells (below).

The presence of subtelomeric sequences at more than one telomere could in principle lead to telomere–telomere exchanges by recombination between non-homologous chromosomes. Such exchanges would provide one explanation for the particular subset of telomeres detected by some of the repeat-containing probes being polymorphic in the population (Brown *et al.* 1990; Cross *et al.* 1990; IJdo *et al.* 1991, 1992*a*). Telomere–telomere exchanges appear to be rare, but one specific example has been described by Wilkie *et al.* (1991). The α-globin gene cluster on human chromosome 16 is found close to the short arm telomere and can be physically linked with sub-terminal sequences. However, there is polymorphism in the population regarding the distance between the α-globin locus and the end of the chromosome. There are three alleles, with the α-globin cluster 170, 350, or 430 kb from the telomere. Two of the alleles correlate with the presence of one particular subtelomeric sequence at 16pter, whereas the third allele is associated with the presence of a different subtelomeric sequence. It has been speculated that this association of different length alleles with different telomere-associated sequences indicates that the observed length polymorphism results from exchanges of long stretches of terminal DNA between non-homologous chromosomes. The point of divergence between the two most common alleles is some 145 kb distal to the α-globin locus. This area has been studied in detail, one feature of note being an unusually large $(CA)_n$ repeat (Wilkie and Higgs 1992). Various lines of evidence suggest that these particular telomere–telomere exchanges did not occur recently and reflect rare events.

The human pseudoautosomal telomere

The human X and Y chromosomes pair in male meiosis in a region of shared homology termed the pseudoautosomal region (Ellis and Goodfellow 1989). This pairing and a postulated obligatory exchange in this region enables these otherwise haploid chromosomes to segregate correctly in meiosis. The region is termed 'pseudoautosomal' because recombination leads to markers not showing strict sex linkage but, instead, simulating autosomal inheritance to varying degrees. There is a gradient of sex linkage in this region such that the marker furthest away from the boundary with the sex-specific regions (DXYS14) recombines between the X and Y chromosomes with a frequency close to 50% in male meiosis. The other loci show a gradient of sex linkage down to 2.5% for *MIC2*, a locus very close to the boundary with the sex-specific regions.

The pseudoautosomal region extends about 2.6 Mb from the Xp and Yp telomeres. The pseudoautosomal telomere has been cloned in yeast (Brown 1989; Brown *et al.* 1990) and long-range restriction maps have been produced by PFGE (Brown 1988; Petit *et al.* 1988; Rappold and Lehrach 1988). The telomeric end is marked by DXYS14, detected by a probe (29c1) that is within 20 kb of the telomere and can be present on a terminal *Bam*HI fragment (Cooke *et al.* 1985; Cooke and Smith 1986). The ability of 29c1 to detect a terminal restriction fragment gave much information on the behaviour of human telomeres at a time before the terminal repeat sequences had been identified. These findings included the heterogeneity of the terminal *Bam*HI fragments, now believed to reflect variation in the size of the (TTAGGG)$_n$ repeat array, and the presence of longer telomeres in sperm when compared with those in lymphocytes (Cooke and Smith 1986; see Chapter 7).

Clustering of human VNTR loci in proterminal regions

The proterminal regions of human chromosomes appear to be significantly enriched for VNTR hypervariable minisatellite loci. Over 70% of hyper-variable minisatellite loci in humans are found at the ends of the linkage maps and by *in situ* hybridization localize to a terminal band (Nakamura *et al.* 1988; Royle *et al.* 1988; Armour *et al.* 1989; Wells *et al.* 1989; Inglehearn and Cooke 1990; Vergnaud *et al.* 1991). The reason for this is not clear and a similar clustering of minisatellite loci has yet to be observed in the mouse (Jeffreys *et al.* 1987; Julier *et al.* 1990; Mariat *et al.* 1993). In contrast to VNTRs, the terminal regions of human chromosomes appear if anything under-represented for microsatellite (e.g. (CA)$_n$ repeat) markers (Weissenbach 1993).

Expansion of the genetic map in telomeric regions of human chromosomes

Seven autosomal telomeres have been placed on both the genetic and physical maps of the human genome; they are the telomeres of 2q, 4p, 7q, 8p, 14q, 16p, and 16q (NIH/CEPH Collaborative Mapping Group, 1992).

On average the genetic map length of a human chromosome measured through female meiosis is some 1.7 times longer than the same chromosome in male meiosis. However, the recombination rate in the terminal regions of the maps are often greater in males than females (for example, see Helms *et al.* 1992). Furthermore, a comparison of the genetic linkage maps with what is currently known of the physical separation of markers on chromosomes suggests that there is expansion of the genetic map (that is, greater recombination for the same physical distance) in the terminal regions of at least some chromosomes for both sexes. The relationship between this terminal map expansion and chiasma distributions is discussed in Chapter 2.

One example of terminal map expansion is provided by chromosome 21. A physical map of around 8 Mb of 21q starting from the telomere, as defined by a subterminal repeat, has been produced (Burmeister *et al.* 1991; Cox and Shimizu 1991; Wang *et al.* 1992) and can be compared with the genetic map of chromosome 21 (Tanzi *et al.* 1988; Warren *et al.* 1989). The total genetic length of 21q is approximately 130 cM. However, the terminal half of 21q22.3, the most terminal band, accounts for around 40% of the genetic length despite being less than 10% of its cytogenetic length. More detailed analysis reveals that this does not reflect a gradual increase in recombination towards the telomere but rather a number of recombination hotspots in the terminal 8 Mb region, including a region within 2 Mb of the telomere where the recombination rate is around five times the average for this chromosome. The terminal map expansion in this example therefore reflects a clustering of hotspots for recombination in subtelomeric regions rather than a progressive increase in recombination frequencies moving towards the telomere.

Another well characterized example is the Xp telomere in female meiosis (Henke *et al.* 1993). Within 100 kb of the terminus there is a recombination hotspot where there is 2.5 cM between two markers no more than 60 kb apart; a genetic distance of < 0.1 cM would have been expected based on the physical distance between the two markers. However, for the remainder of the 1.8 Mb subterminal region analysed the total genetic length (2.7 cM) is comparable to the recombination rate for the X chromosome as a whole in female meiosis (164 Mb, 208 cM). There is no dramatic map expansion at this telomere in females, and the localized hotspot probably reflects primary sequence rather than telomeric location.

Too much weight should not be placed on such studies at this time, as they have a rather coarse level of resolution. However, as the human genome mapping project continues we will soon arrive at a more detailed understanding of the relationship between genetic and physical distances in different regions of the human genome.

Telomere-associated chromosome fragmentation in mammalian cells

Until recently a major problem in the analysis of mammalian telomeres was the lack of a method to alter telomere structure in the cell. In yeast the structure of endogenous chromosome ends can be altered by homologous recombination, or a linear minichromosome can be introduced that terminates in any desired sequence, and the behaviour of the novel chromosome can then be followed in mitosis and meiosis (Vollrath *et al.* 1988; Pavan *et al.* 1990). However, work now indicates that mammalian telomeres can be functionally reintroduced into cells in culture. Farr *et al.* (1991) transformed a Chinese hamster cell line in culture with a linear DNA molecule carrying a selectable marker and terminating in $(TTAGGG)_n$ repeats and analysed random integration events. In around one-third of the cell lines the transfected DNA was telomeric, as defined by both *in situ* hybridization and exonuclease sensitivity. Furthermore, in any one cell line the terminal restriction fragments were heterogeneous in size and larger than the equivalent fragment of the introduced DNA. A similar random breakage approach in human and mouse cells has been reported by Barnett *et al.* (1993). In both cases it seems likely that integration at an interstitial site has been accompanied by chromosome breakage, with the introduced (TTAGGG)$_n$ repeats both stabilizing the end of the chromosome and providing a substrate for further (TTAGGG)$_n$ repeat addition by telomerase. An important biological conclusion from these studies is that (TTAGGG)$_n$ repeats, by the limited criterion of stabilizing the end of a broken chromosome in a cultured cell, appear sufficient for telomere function.

Telomere-associated chromosome fragmentation has been extended by combining it with homologous recombination. By targeting a gene on chromosome 1 with a telomere-containing construct, a chromosome broken at the correct site was obtained (Itzhaki *et al.* 1992). This will be a powerful new way to produce large, defined changes in mammalian chromosome structure. Telomere-associated chromosome fragmentation also provides a new technique to map complex genomes. By selecting for loss of a distal marker, Farr *et al.* (1992) have created a panel of somatic cell hybrids with nested terminal deletions of the long arm of the human X chromosome. Markers can be quickly ordered based on their presence or absence in the various derivative X chromosome lines. The result is a powerful and rapid way to map markers at a fairly coarse level of resolution. Unfortunately, for the human X chromosome these panels will probably soon become redundant with the advent of ordered YAC and cosmid contigs from the mapping projects, for example as are already available for the human Y chromosome. However, this approach may well be of considerable use in the mapping of less extensively studied genomes, such as those of commercially important animals like the pig.

Some internal (TTAGGG)$_n$ arrays may be the remnants of telomere fusion

Based on similarities in banding patterns and hybridization homologies it has been suggested that human chromosome 2 arose from the fusion of two great ape chromosomes, thereby reducing the chromosome number from 24 to 23 pairs in humans (Yunis and Prakash 1982). The cytogenetic data point to a fusion in the region of 2q1, and *in situ* hybridization with both (TTGGGG)$_n$ (Allshire *et al.* 1988) and a probe containing degenerate (TTAGGG)$_n$ repeats (Wells *et al.* 1990) detects telomere-like sequences in the vicinity of 2q13. This fusion site has been cloned in cosmids and sequenced, revealing two degenerate (TTAGGG)$_n$ arrays in a head-to-head arrangement (IJdo *et al.* 1991). Flanking this inverted repeat are sequences over 90% identical to known subtelomeric sequences, and indeed *in situ* hybridization probes containing these flanking sequences detect some chromosome termini as well as 2q13. Conversely, a subtelomeric sequence isolated from a cloned telomere by Cross *et al.* (1990) hybridizes to a number of telomeres and also 2q13. In short, the molecular structure of this region of 2q13 agrees well with what might be predicted from a telomere–telomere fusion of two chromosomes. Such a fusion would presumably have initially been dicentric and the small amount of α-satellite sequence detectable at 2q21.3–2q22.1 may be the remnant of this second centromere (Avarello *et al.* 1992).

The formation of human chromosome 2 is the only obvious telomere–telomere fusion that has occurred with respect to the great ape genomes. In addition to this locus there are other small internal (TTAGGG)$_n$ tracts in the human genome, some of which have been isolated by PCR (Weber *et al.* 1990, 1991*a*). Two such internal (TTAGGG)$_n$ tracts have been mapped to interstitial locations (8p21 and 10q22–10q24) using cosmids containing them (Weber *et al.* 1991*a*). They could in principle be the remnants of karyotypic changes from earlier in vertebrate evolution, although there is no evidence as yet to support this suggestion.

Using telomeric sequences to investigate chromosome structure

The availability of *in situ* probes for telomeres has enabled a range of questions regarding chromosome structure to be addressed. For example, a number of individuals with ring chromosomes have been shown to have (TTAGGG)$_n$ and subtelomeric sequences at the fusion site (Park *et al.* 1992; Pezzolo *et al.* 1993). The evidence from high resolution chromosome banding is that there are no detectable deleted regions on these chromosomes (Pezzolo *et al.* 1993; Sawyer *et al.* 1993). If, as is possible, these particular ring chromosomes have been formed by telomere–telomere fusion with no major loss of chromosomal material, why do patients carrying such chromosomes have the clinical symptoms (failure to thrive and minor dysmorphic signs) of what is termed 'ring syndrome'? One speculation is that ring chromosomes are more unstable than an equivalent linear chromosome because sister chromatid exchange can lead to the formation of a dicentric chromosome that will be highly prone to loss or rearrangement (Fig. 1.7). The loss or rearrangement of this ring chromosome in some somatic cells provides one possible cause of the clinical symptoms.

Another question is the structure of double minute chromosomes, which are common products of gene amplification events. Electron microscopy and the inability of double minute chromosomes to migrate into pulsed field gels unless X-irradiated suggests they are circular structures. The larger, cytologically visible double minutes are not amenable to such an approach, but it has been shown by *in situ* hybridization that they do not carry (TTAGGG)$_n$ sequences. Assuming (TTAGGG)$_n$ is necessary for mammalian telomere function this argues that these larger double minutes are also circular in form (Lin *et al.* 1990; Furuya *et al.* 1993).

When using (TTAGGG)$_n$ as a probe for *in situ* hybridization the signal is often seen at the very tip of prophase chromosomes. As the chromosomes condense the location of the signal can appear to change, so that with metaphase chromosomes there can be noticeable DNA counter-staining beyond the site of hybridization (see for example Moyzis *et al*. 1988). Moens and Pearlman (1990) have shown that mouse telomeres are associated with the cores of meiotic chromosomes, with loops of DNA extending out and beyond the telomeres. It is possible to speculate that the DNA counter-staining seen beyond the (TTAGGG)$_n$ signal corresponds to scaffold-attached loops.

Cryptic translocations and human genetic disease

A translocation can usually be identified by conventional chromosome banding techniques. However, if the amount of material transferred is very small it may not be detectable; this is termed cryptic translocation (reviewed by Ledbetter 1992). Such translocations can, however, be detected by the use of molecular markers for the terminal chromosomal regions, and the following examples illustrate the degree to which such molecular probes can provide clinicians with a greater level of resolution than that provided by banded chromosomes.

Lamb *et al*. (1989) have described an α-thalassaemia patient with an apparently normal karyotype as judged by conventional chromosome banding techniques. Molecular probes revealed a cryptic translocation between the tips of 1p and 16p. This was balanced in the mother, but the chromosomes segregated so that the affected son was trisomic for 1pter and monosomic for 16pter, the latter causing the α-thalassaemia. Such analysis can also detect terminal deletions of 16p for chromosomes that appear normal cytogenetically (Wilkie *et al*. 1990). One particular example removes 2 Mb from 16pter and the break is healed by addition of TTAGGG repeats (Lamb *et al*. 1993; see Chapter 6).

Overhauser *et al*. (1989) have described a patient with cri-du-chat (5p$^-$) syndrome. Family members had apparently normal karyotypes by conventional chromosome banding, but the use of molecular markers for distal 5p revealed a submicroscopic balanced translocation in one parent. Bernstein *et al*. (1993) have detected a cryptic balanced translocation involving 5p in the mother of another cri-du-chat case by *in situ* hybridization using a chromosome 5 specific probe.

Altherr *et al*. (1991) have described a patient with Wolf–Hirschhorn syndrome, which usually results from deletion or translocation of material from distal 4p. This particular patient showed no obvious karyotypic change as judged by conventional banding. However, by performing *in situ* hybridization with probes for the telomeric region of 4p it was shown that the patient's mother carried a submicroscopic balanced t(4; 19) translocation, leading to the patient being hemizygous for distal 4p.

Kuwano *et al.* (1991) have reported a patient with Miller–Dieker syndrome, for which the critical region is 17p13.3. This patient has an apparently normal karyotype, but the use of *in situ* probes for distal 17p revealed a cryptic t(8q; 17p) balanced translocation in the father, leading to hemizygosity for the tip of 17p in the affected offspring.

Meltzer *et al.* (1993) have analysed a number of melanoma tumour biopsies and cell lines with chromosomes showing apparent terminal deletions. Using *in situ* hybridization probes obtained by microdissecting the ends of the 'terminally deleted' chromosomes they demonstrate that several examples are in fact subtelomeric translocations; the 'deletion' chromosome terminates in a submicroscopic amount of telomeric sequence originating from another chromosome. It is possible that other examples of apparent terminal deletion are also cryptic translocations.

Finally, Migeon *et al.* (1993) report a cryptic translocation of a region of Xq28 to an autosome resulting in haemophilia A, again with no visible change in Xq28.

Cytogenetically invisible subtelomeric translocations may be a more important cause of genetic disease than previously thought. One reason for this is that the amount of deleted material is small and therefore unbalanced offspring are more likely to survive than would be the case if they had visible (and therefore much larger) unbalanced translocations. For example, the terminal deletions associated with α-thalassaemia described here and in Chapter 6 involve a locus very close to the telomere. Relatively small amounts of chromosomal material are removed in these terminal deletions, from 300 kb to around 2000 kb, and clearly these deletions do not remove any distal genes for which haplo-insufficiency causes lethality. Other potential candidates for terminal deletions healed by $(TTAGGG)_n$ addition would include some examples of Wolf–Hirschhorn, Miller–Dieker, and cri-du-chat syndromes, although this has not been proven as yet.

The potential clinical significance of cryptic translocations and terminal deletions is such that it is important to consider specific approaches to detect them. Molecular markers for distal regions can be used for both DNA analysis and *in situ* hybridization. The high density of VNTR loci in terminal regions has been proposed as the basis of a strategy to look for unbalanced translocations in patients with unexplained mental retardation (Wilkie 1993).

Fragile sites and telomeres

Fragile sites are a cytological phenomenon, manifesting themselves as microscopically visible breaks or gaps in a chromosome (reviewed by Sutherland and Hecht 1985). They fall into two classes: common, which are found in all or most individuals, and rare, which are found only in certain individuals. Most have no obvious clinical symptoms, although

FRAXA (below) is associated with mental retardation. To visualize such breaks at a reasonable frequency requires cells to be cultured in special media which induce their expression. What these media have in common is that they perturb DNA replication either directly (e.g. aphidicolin) or by altering intracellular nucleotide pools (e.g. folate starvation). This has led to the suggestion that fragile sites reflect problems of replicating some regions of the genome under certain conditions (Laird *et al.* 1987; Sutherland 1988) although it remains unclear how such replication problems might produce the cytogenetic phenotype.

Two observations led to the speculation that some fragile sites might be caused by interstitial telomeric sequences (Hastie and Allshire 1989). Firstly, in yeast a head-to-head inverted repeat of *Tetrahymena* telomere repeats resolves to give two functional telomeres at a frequency of around 10^{-2} per cell division (Murray *et al.* 1988). A similar resolution of interstitial telomeric repeats in human chromosomes might in theory produce a cytogenetic phenotype similar to a fragile site. The resolution mechanism is not understood, but one might speculate that it could reflect the ability of such an inverted repeat array to form a cruciform structure that might then be cleaved by a recombinase specific for such structures (see Fig. 1.2). Such 'Holliday resolvases' have been identified in a number of species (see Chapter 1). The free ends produced by such a cleavage could then seed the addition of yeast TG_{1-3} repeats, resulting in the observed linearization reaction. It is important to ask whether an inverted repeat array of $(TTGGGG)_n$ sequences is more susceptible to cleavage than a perfect inverted repeat of a non-telomeric sequence, and in fact there is little evidence that would support such a suggestion. Palindromes are genetically unstable in yeast (Gordenin *et al.* 1993; Henderson and Petes 1993; Ruskin and Fink 1993). It is therefore possible to speculate that success in detecting cleavage of *Tetrahymena* repeats reflects not so much a specific instability of telomere-like sequences but rather the ability of the cleavage product to be healed by TG_{1-3} repeat addition and thus give a detectable product.

The second observation that suggested telomeric sequences might have an effect on metaphase chromosome structure came from the analysis of a patient with a t(6; 19) translocation (Drets and Therman 1983). This chromosome appears to be the result of a telomere–telomere fusion, as judged by chromosome banding. In some cells one or both chromatids were apparently broken at the fusion point, at a frequency much greater than for any other region of the genome. It is difficult to draw any firm conclusions from this study, especially as the molecular structure of this locus and in particular the presence of any telomeric sequences is not known. However, fragility at the site of an apparent telomere–telomere fusion did suggest a link between telomeres and fragility.

The best characterized human fragile site is FRAXA at Xq27, as its expression is associated with a severe and common mental retardation syndrome in males (reviewed by Oostra and Verkerk 1992; Trottier *et al.* 1993). The chromosome breaks seen cytologically occur in the vicinity of an amplified $(CCG)_n$ repeat. Normal individuals have about 100 bp of the repeat at this site, whereas affected males have from 500 to over 5000 bp of $(CCG)_n$. The expansion of this repeat is thought to cause clinical symptoms via an effect on the *FMR-1* gene at this site. This mutation also has a visible cytological phenotype, the fragile site. By *in situ* hybridization to chromosomes expressing the fragile site, it has been shown that flanking sequences within a kilobase of the $(CCG)_n$ repeat hybridize to opposite sides of the gap, arguing that the break occurs within the $(CCG)_n$ array itself (Kremer *et al.* 1991). FRAXA requires induction by medium with an excess of thymidine, which is thought to inhibit ribonucleotide reductase and result in a deficiency of intracellular dCTP for DNA synthesis. One speculation is that under such conditions it might be difficult to replicate $(CCG)_n$ arrays, resulting in single-stranded DNA regions that could disrupt chromosome condensation and manifest themselves as a fragile site. Whatever the explanation, the main conclusion from this analysis is that FRAXA is clearly not associated with an internal $(TTAGGG)_n$ array. FRAXE is another fragile site on the X chromosome, located close to but distinct from FRAXA, and has been analysed recently and shown to contain a similarly expanded $(CCG)_n$ array in patients expressing the fragile site (Knight *et al.* 1993). In contrast expanded $(CAG)_n$ arrays, such as in myotonic dystrophy (below), do not create fragile sites.

Another fragile site that has been analysed in some detail is FRA2B, a rare folate-sensitive fragile site at 2q13. This is close to the ancestral fusion site at 2q13 where there is an inverted head-to-head $(TTAGGG)_n$ array (see above). However, a cosmid spanning the fusion point maps by *in situ* hybridization to the distal side of the fragile site (IJdo *et al.*, 1992*b*). Thus FRA2B also does not appear to be associated with an internal telomere sequence array.

There are additional fragile sites under analysis. For example, one cosmid which contains subtelomeric sequences also detects 3p14, a band which contains a common fragile site (IJdo *et al.* 1991). A second cosmid containing subtelomeric sequences that hybridizes to many telomeres also hybridizes to a single interstitial site at 13q21.3–q22, another region containing a common fragile site (IJdo *et al.* 1992*a*).

Although the evidence described here suggests that the fragile sites analysed so far do not correlate with internal telomeric sequences, it is important to note that fragile sites can behave differently, such as in the spectrum of drugs which induce their expression (Sutherland and Hecht 1985). This variation in phenotype suggests there may be variation in the DNA sequence underlying different fragile sites, and it remains a formal possibility that some of the uncharacterized fragile sites correspond to internal telomere tracts.

Other cytological abnormalities possibly involving telomeres

Park *et al.* (1992) and Rossi *et al.* (1993) have reported a number of constitutional chromosome abnormalities where a translocation has occurred involving a terminal band. A number of examples were found to possess $(TTAGGG)_n$ at the translocation breakpoint as judged by *in situ* hybridization. The simplest explanation of this observation is that a fusion to the intact terminus of the recipient chromosome has occurred, although this remains speculative in the absence of further molecular details.

Another phenomenon involving telomeres is that of 'jumping translocations'. For example, in very rare cases a malignancy will show that a chromosome segment has 'jumped' to a number of different telomeric locations in different cells (Shippey *et al.* 1990). As judged by chromosome banding the extra material appears to have been added directly onto the end of the recipient chromosome as if it were a fusion with the intact end. One example of a jumping translocation has been shown to retain $(TTAGGG)_n$ at the fusion points. This is a Prader–Willi patient showing mosaicism, with jumping translocations of 15q13–qter to a number of different telomeres (Park *et al.* 1992). Why these rare cases seem to permit numerous translocations involving telomeres is completely unknown, but as it might reflect some change in telomere metabolism allowing fusions to occur it would be interesting to investigate this behaviour further. Another Prader–Willi case with an unbalanced translocation of a region of 15q onto the end of chromosome 12 also has $(TTAGGG)_n$ at the translocation breakpoint (Reeve *et al.* 1993). One important potential conclusion from the presence of $(TTAGGG)_n$ at sites of apparent telomere–telomere fusion in jumping translocations and the other chromosome abnormalities described above, as well as the telomere–telomere fusion that occurred to form human chromosome 2, is that telomere–telomere fusions may occasionally occur despite one or both telomeres terminating in $(TTAGGG)_n$. Although it may happen only rarely, it seems possible that human telomeric sequences do not protect completely against end-to-end fusion.

Comparative structure of mouse telomeres

The structure of the mouse telomere has yet to be described with the same level of detail as that of the human telomere. Strong evidence for the mouse terminal repeat sequence being TTAGGG is the identification of a telomerase activity from mouse cells which adds this sequence (Chapter 4). $(TTAGGG)_n$ probes produce a telomeric signal on metaphase mouse chromosomes (Meyne *et al.* 1989) and most of the $(TTAGGG)_n$ signal is BAL31-sensitive on Southern blots (Kipling and Cooke 1990; Starling *et al.* 1990). Where the structure of mouse telomeres differ most strikingly from those of humans is that their terminal repeat arrays are much larger.

Following digestion with enzymes with a 4 bp recognition sequence, (TTAGGG)$_n$-containing fragments as large as 150 kb can be detected using PFGE. The length heterogeneity of any one telomere is sufficiently low that the signal is not the smear as seen for human DNA, but is rather a series of incompletely resolved fragments (Kipling and Cooke 1990; Starling *et al.* 1990). The size range of fragments produced depends on the mouse strain used. For example, DBA/2 fragments are in the 20–150 kb range, whereas those of C57BL/6 are in a much tighter range (20–65 kb). This suggests that average telomere length is influenced by the genotype of the mouse; examples of this phenomenon in other species are given in Chapter 5, where the genetic control of telomere length is discussed.

The ability to visualize mouse telomeres as discrete restriction fragments has revealed a marked genetic instability in the length of the terminal repeat arrays. Even for highly inbred mice, the size of (TTAGGG)$_n$-hybridizing fragments varies from mouse to mouse. This hypervariability appears to reflect a high rate of generation of alleles of new size. In family studies most fragments do behave in a straightforward Mendelian fashion, but new size alleles can be seen in some offspring (Kipling and Cooke 1990; Starling *et al.* 1990). The mechanism underlying the generation of alleles of new size remains a mystery. One possibility is that it reflects turnover of telomeric sequences, creating heterogeneity in germ-line cells and thus in the mouse that is derived from a single germ-line cell. This would be analogous to experiments in *S. cerevisiae* where individual cells have been cloned out of a population and shown to have different telomere lengths (Shampay and Blackburn 1988). One argument against this model is that it would be predicted to affect all telomeres equally, whereas the situation in mice is that most fragments are the same length in parents and offspring; the new size alleles are the result of one or two telomeres undergoing a marked change in length, in some cases of more than 20 kb. This argues that simple turnover of terminal repeat sequences and selection out of the resulting heterogeneous population of cells is not sufficient to explain the large size changes seen. Although the mechanism is not understood, one speculation is that the large size changes reflect frequent telomere–telomere exchanges.

Mouse subtelomeric regions are poorly characterized, although Broccoli *et al.* (1992) have isolated a (TTAGGG)$_n$-adjacent sequence by a PCR approach. The *in situ* hybridization evidence is consistent with it being present at most if not all telomeres, although it is not known if this sequence is BAL31-sensitive on Southern blots. One sequence that is clearly sub-terminal is the mouse minor satellite (Kipling *et al.* 1991). This is found as a tandem repeat array composed of underlying 120 bp monomer units. Around 300 kb of this sequence is found at the terminally located centro-mere of each *Mus musculus* chromosome, with the exception of the Y. With the α-satellite that is found at human centromeres, it shares a 17 bp motif that binds the centromere-associated CENP-B protein, and the minor satellite may have a role in centromere function for this reason. The minor satellite arrays are so close to the end of the chromosome that they can be on terminal restriction fragments shorter than 1 Mb that also contain $(TTAGGG)_n$. As enzymes such as *Bgl*II, *Pvu*II, and *Hind*III generally do not cleave between the minor satellite and $(TTAGGG)_n$ this argues that there is unlikely to be much in the way of potentially gene-containing sequence between these two sequences; mouse chromosomes seem telocentric. Curiously, *Xba*I and *Bam*HI inevitably do cleave between the minor satellite and the telomere. It has been speculated that one explanation is a subtelomeric sequence, containing or lacking the appropriate sites, that is shared by most of the centromere-proximal telomeres thus resulting in a shared terminal restriction map (Kipling *et al.* 1991). This would be a similar situation to that found in humans, where shared subtelomeric sequences result in similarities in the terminal restriction maps (Brown *et al.* 1990; Cross *et al.* 1990; Cheng *et al.* 1991). However, such subtelomeric sequences have yet to be isolated in the mouse.

There are some interstitial, BAL31-insensitive $(TTAGGG)_n$ tracts in the mouse, and a 300 bp clone with two or three telomere repeats in an inverted orientation at either end of a low copy number sequence has been isolated by PCR (Bouffler *et al.* 1993*b*). This clone is being investigated with respect to a hypothesis that interstitial telomere-like sequences are involved in the susceptibility of CBA/H mice to irradiation-induced acute myeloid leukaemia (Bouffler *et al.* 1993*a,b*; Silver and Cox 1993).

Genetic analysis of mouse telomeres: closing the genetic map

Polymorphic $(TTAGGG)_n$-containing fragments have been identified by conventional gel electrophoresis. Some of these have been mapped in recombinant inbred strains and correspond to the distal ends of the sex chromosomes (that is, the pseudoautosomal region; Eicher *et al.* 1992), the distal telomeres of chromosomes 4 and 9 (Elliott and Yen 1991), and the proximal telomere of the Y chromosome (Elliott and Yen 1991; Eicher *et al.* 1992). In a similar fashion $(TTAGGG)_n$ polymorphisms have been mapped in an interspecific backcross panel to the distal ends of *Mus spretus* chromosome 13 and the pseudoautosomal region of the sex chromosomes (Eicher and Shown 1993). The size of a 1.3 Mb $(TTAGGG)_n$-containing terminal restriction fragment from the centromere-proximal telomere of chromosome 7 differs between C57BL/6 and DBA/2, probably due to differences in the amount of centromeric satellite sequence included on the fragment (Kipling *et al.* 1991). This polymorphism has been mapped in the BXD recombinant inbred strains to a position consistent with the centromere-proximal telomere of chromosome 7 (Kipling *et al.* 1991).

Perspective: (TTAGGG)$_n$ tract instability and human genetic disease

The mouse provides a dramatic example of instability in terminal repeat array length, which has been speculated to reflect more than telomerase-mediated turnover of sequence. It is possible that this instability has some mechanistic similarity to the instability of short tandem repeats that produce variable microsatellite and minisatellite loci, and the instability of trinucleotide repeats that underlie some human genetic diseases. The mechanism responsible for the instability of short tandem repeats remains controversial. One model of minisatellite evolution invokes slippage on repetitive templates during DNA replication; short length changes might be caused by strand slippage during replication (Levinson and Gutman 1987). Another possible mechanism is unequal sister chromatid exchange. A third suggestion is that it involves unequal exchange between homologues during meiosis or mitosis. This model has been tested experimentally as such an event would lead to markers flanking the locus being recombinant following the mutation.

Studies of both somatic and meiotic mutations of mouse and human minisatellite loci indicate that flanking marker exchange does not occur with mutation (Wolff *et al.* 1988, 1989; Mahtani and Willard 1993). For example, Kelly *et al.* (1991) have analysed both somatic and germ-line mutations at the mouse hypervariable minisatellite locus *Ms6-hm*. This locus is very unstable and shows new length alleles equivalent to a gamete mutation rate of 2.5%. The new mutations, which can also occur in somatic tissue, cause a change in the length of its large (< 6 kb) (GGGCA)$_n$ repeat array and are not associated with exchange of flanking markers. Similarly, detailed sequence analysis of the human minisatellite locus *MS3* indicates that new alleles arise along haploid chromosome lineages, with no evidence for exchange between homologues (Jeffreys *et al.* 1990). Repeat units of the human minisatellite *MS1* can be lost and gained at high frequency even when integrated into a yeast chromosome in a haploid strain, which has no homologous chromosome to exchange with (Cederberg *et al.* 1993). This also demonstrates that the instability of these loci is likely to be a fundamental property of the underlying DNA sequence in a eukaryotic cell and not a peculiarity of mammals.

Another example of the instability of short tandem repeats is that of microsatellite loci, such as the (CA)$_n$ repeats that have been a source of plentiful polymorphic markers for mapping mammalian genomes (Weissenbach 1993). (CA)$_n$ repeats show *RAD52*-independent instability in yeast (Henderson and Petes 1992). *RAD52* is required for most types of mitotic recombination and this therefore argues against recombination-based models of instability. However, not all recombination is abolished in *rad52* strains; for example, unequal sister chromatid exchange within the rDNA locus is unaffected by a *rad52* mutation. (CA)$_n$ repeat instability can occur in mammalian cells in culture (Farber *et al.* 1994). Genetic analysis of a $(CA)_n$ repeat in the cystic fibrosis transmembrane conductance regulator gene argues that unequal exchange between homologous chromosomes is not responsible for the creation of new alleles at microsatellite loci (Morral *et al.* 1991). A large study of microsatellite mutations in the CEPH families concluded that none of the mutation events occurred at the sites of meiotic recombination (Weber and Wong 1993). Unequal crossover between homologous chromosomes does not therefore appear to be a candidate for the formation of the majority of new micro- and minisatellite mutations. Replication can slip on $(CA)_n$ and other short repeat arrays *in vitro*, leading to incorrect amounts of the complementary strand being synthesized (Chapter 1; Schlötterer and Tautz 1992), and such slippage synthesis is a strong candidate for producing new microsatellite alleles.

The instability of short tandem repeats can have clinical consequences (Richards and Sutherland 1992; Kuhl and Caskey 1993). The type of mutation underlying a number of genetic diseases has now been identified as changes in the length of a trinucleotide repeat array at the disease locus. For example, amplification of a $(CCG)_n$ repeat array is involved in fragile X syndrome (see above). Similarly, amplification of $(CAG)_n$ repeat arrays underlie myotonic dystrophy, spinobulbar muscular atrophy, spinocerebellar ataxia type 1, and Huntingdon's disease (Richards and Sutherland 1992; Kuhl and Caskey 1993; Orr *et al.* 1993). Trinucleotide repeat instability can manifest itself in both meiosis and mitosis, with some patients carrying the most unstable length alleles showing somatic variation in array size (Kuhl and Caskey 1993). Understanding the mechanism whereby these mutations arise is a major goal.

References

Reviews

Bouffler, S., Silver, A., and Cox, R. (1993*a*). The role of DNA repeats and associated secondary structures in genomic instability and neoplasia. *Bioessays*, **15**, 409–12.

Broccoli, D. and Cooke, H. (1993). Aging, healing, and the metabolism of telomeres. *Am. J. Human Genet.*, **52**, 657–60.

Ellis, N. and Goodfellow, P. N. (1989). The mammalian pseudoautosomal region. *Trends Genet.*, **5**, 406–10.

Hastie, N. D. and Allshire, R. C. (1989). Human telomeres: fusion and interstitial sites. *Trends Genet.*, **5**, 326–30.

Kipling, D. and Cooke, H. J. (1992). Beginning or end? Telomere structure, genetics and biology. *Human Mol. Genet.*, **1**, 3–6.

Kuhl, D. P. A. and Caskey, C. T. (1993). Trinucleotide repeats and genome variation. *Curr. Opin. Genet. Dev.*, **3**, 404–7.

Laird, C., Jaffe, E., Karpen, G., Lamb, M., and Nelson, R. (1987). Fragile sites in human chromosomes as regions of late-replicating DNA. *Trends Genet.*, **3**, 274-81.

Ledbetter, D. H. (1992). Cryptic translocations and telomere integrity. *Am. J. Human Genet.*, **51**, 451-6.

Oostra, B. A. and Verkerk, A. J. M. H. (1992). The fragile X syndrome: isolation of the FMR-1 gene and characterization of the fragile X mutation. *Chromosoma*, **101**, 381-7.

Richards, R. I. and Sutherland, G. R. (1992). Dynamic mutations: a new class of mutations causing human disease. *Cell*, **70**, 709-12.

Sutherland, G. R. (1988). The role of nucleotides in human fragile site expression. *Mutat. Res.*, **200**, 207-13.

Sutherland, G. R. and Hecht, F. (1985). *Fragile sites on human chromosomes*. Oxford University Press, Oxford.

Trottier, Y., Devys, D., and Mandel, J. L. (1993). An expanding story. *Curr. Biol.*, **3**, 783-6.

Weissenbach, J. (1993). Microsatellite polymorphisms and the genetic linkage map of the human genome. *Curr. Opin. Genet. Dev.*, **3**, 414-17.

Primary papers

Allshire, R. C., Gosden, J. R., Cross, S. H., Cranston, G., Rout, D., Sugawara, N., *et al.* (1988). Telomeric repeat from *T. thermophila* cross hybridizes with human telomeres. *Nature*, **332**, 656-9.

Allshire, R. C., Dempster, M., and Hastie, N. D. (1989). Human telomeres contain at least three types of G-rich repeat distributed non-randomly. *Nucleic Acids Res.*, **17**, 4611-27.

Altherr, M. R., Bengtsson, U., Elder, F. F. B., Ledbetter, D. H., Wasmuth, J. J., McDonald, M. E., *et al.* (1991). Molecular confirmation of Wolf-Hirschhorn syndrome with a subtle translocation of chromosome 4. *Am. J. Human Genet.*, **49**, 1235-42.

Armour, J. A. L., Wong, Z., Wilson, V., Royle, N. J. and Jeffreys, A. J. (1989). Sequences flanking the repeat arrays of human minisatellites: association with tandem and dispersed repeat elements. *Nucleic Acids Res.*, **17**, 4925-35.

Avarello, R., Pedicini, A., Caiulo, A., Zuffardi, O., and Fraccaro, M. (1992). Evidence for an ancestral alphoid domain on the long arm of human chromosome 2. *Human Genet.*, **89**, 247-9.

Barnett, M. A., Buckle, V. J., Evans, E. P., Porter, A. C. G., Rout D., Smith, A. G., and Brown, W. R. A. (1993). Telomere directed fragmentation of mammalian chromosomes. *Nucleic Acids Res.*, **21**, 27-36.

Bates, G. P., MacDonald, M. E., Baxendale, S., Sedlacek, Z., Youngman, S., Romano, D., *et al.* (1990). A yeast artificial chromosome telomere clone spanning a possible location of the Huntingdon disease gene. *Am. J. Human Genet.*, **46**, 762-75.

Bernstein, R., Bocian, M. E., Cain, M. J., Bengtsson, U., and Wasmuth, J. J. (1993). Identification of a cryptic t(5; 7) reciprocal translocation by fluorescent in situ hybridization. *Am. J. Med. Genet.*, **46**, 77-82.

Bouffler, S., Silver, A., Papworth, D., Coates, J., and Cox, R. (1993*b*). Murine radiation myeloid leukaemogenesis: relationship between interstitial telomere-like sequences and chromosome 2 fragile sites. *Genes Chromosomes Cancer*, **6**, 98-106.

Broccoli, D., Miller, O. J., and Miller, D. A. (1992). Isolation and characterization of a mouse subtelomeric sequence. *Chromosoma*, **101**, 442–7.

Brown, W. R. A. (1988). A physical map of the human pseudoautosomal region. *EMBO J.*, **7**, 2377–85.

Brown, W. R. A. (1989). Molecular cloning of human telomeres in yeast. *Nature*, **338**, 774–6.

Brown, W. R. A., MacKinnon, P. J., Villasanté, A., Spurr, N., Buckle, V. J., and Dobson, M. J. (1990). Structure and polymorphism of human telomere-associated DNA. *Cell*, **63**, 119–32.

Burmeister, K., Kim, S., Price, E. R., de Lange, T., Tantravahi, U., Myers, R. M., and Cox, D. R. (1991). A map of the distal region of the long arm of human chromosome 21 constructed by radiation hybrid mapping and pulsed-field gel electrophoresis. *Genomics*, **9**, 19–30.

Cederberg, H., Agurell, E., Hedenskog, M., and Rannug, U. (1993). Amplification and loss of repeat units of the human minisatellite MS1 integrated in chromosome III of a haploid yeast strain. *Mol. Gen. Genet.*, **238**, 38–42.

Cheng, J. -F., Smith, C. L., and Cantor, C. R. (1989). Isolation and characterization of a human telomere. *Nucleic Acids Res.*, **17**, 6109–27.

Cheng, J. -F., Smith, C. L. and Cantor, C. R. (1991) Structural and transcriptional analysis of a human subtelomeric repeat. *Nucleic Acids Res.*, **19**, 149–54.

Cooke, H. J. and Smith, B. A. (1986). Variability at the telomeres of the human X/Y pseudoautosomal region. *Cold Spring Harbor Symp. Quant. Biol.*, **51**, 213–19.

Cooke, H. J., Brown, W. R. A., and Rappold, G. A. (1985). Hypervariable telomeric sequences from the human sex chromosomes are pseudoautosomal. *Nature*, **317**, 687–92.

Cox, D. R. and Shimizu, N. (1991). Report of the committee on the genetic constitution of chromosome 21. *Cytogenet. Cell Genet.*, **58**, 800–26.

Cross, S. H., Allshire, R. C., McKay, S. J., McGill, N. I., and Cooke, H. J. (1989). Cloning of human telomeres by complementation in yeast. *Nature*, **338**, 771–4.

Cross, S., Lindsey, J., Fantes, J., McKay, S., McGill, N., and Cooke, H. (1990). The structure of a subterminal repeated sequence present on many human chromosomes. *Nucleic Acids Res.*, **18**, 6649–57.

de Lange, T., Shiue, L., Myers, R. M., Cox, D. R., Naylor, S. L., Killery, A. M., and Varmus, H. E. (1990). Structure and variability of human chromosome ends. *Mol. Cell. Biol.*, **10**, 518–27.

Dietz-Band, J., Riethman, H., Hildebrand, C. E., and Moyzis, R. (1990). Characterization of polymorphic loci on a telomeric fragment of DNA from the long arm of human chromosome 7. *Genomics*, **8**, 168–70.

Drets, M. E. and Therman, E. (1983). Human telomeric 6; 19 translocation chromosome with a tendency to break at the fusion point. *Chromosoma*, **88**, 139–44.

Eicher, E. M. and Shown, E. P. (1993). Molecular markers that define the distal ends of mouse autosomes 4, 13, and 19 and the sex chromosomes. *Mammal. Genome*, **4**, 226–9.

Eicher, E. M., Lee, B. K., Washburn, L. L., Hale, D. W., and King, T. R. (1992). Telomere-related markers for the pseudoautosomal region of the mouse genome. *Proc. Natl. Acad. Sci. USA*, **89**, 2160–4.

Elliott, R. W. and Yen, C. H. (1991). DNA variants with telomere probe enable genetic mapping of ends of mouse chromosomes. *Mammal. Genome*, **1**, 118–22.

Farber, R. A., Petes, T. D., Dominska, M., Hudgens, S. S., and Liskay, R. M.

(1994). Instability of simple sequence repeats in a mammalian cell line. *Human Mol. Genet.*, **3**, 253-6.

Farr, C., Fantes, J., Goodfellow, P., and Cooke, H. (1991). Functional reintroduction of human telomeres into mammalian cells. *Proc. Natl. Acad. Sci. USA*, **88**, 7006-10.

Farr, C. J., Stevanovic, M., Thomson, E. J., Goodfellow, P. N., and Cooke, H. J. (1992). Telomere-associated chromosome fragmentation: applications in genome manipulation and analysis. *Nature Genet.*, **2**, 275-82.

Fry, K. and Salser, W. (1977). Nucleotide sequences of HS-α satellite DNA from kangaroo rat Dipodomys ordii and characterization of similar sequences in other rodents. *Cell*, **12**, 1069-84.

Furuya, T., Morgan, R., Berger, C. S., and Sandberg, A. A. (1993). Presence of telomeric sequences on deleted chromosomes and their absence on double minutes in cell line HL-60. *Cancer Genet. Cytogenet.*, **70**, 132-5.

Gordenin, D. A., Lobachev, K. S., Degtyareva, N. P., Malkova, A. L., Perkins, E., and Resnick, M. A. (1993). Inverted DNA repeats: a source of eukaryotic genomic instability. *Mol. Cell. Biol.*, **13**, 5315-22.

Guerrini, A. M., Ascenzioni, F., Pisani, G., Rappazzo, G., Valle, G. D., and Donini, P. (1990). Cloning a fragment from the telomere of the long arm of human chromosome 9 in a YAC vector. *Chromosoma*, **99**, 138-42.

Hastie, N. D., Dempster, M., Dunlop, M. G., Thompson, A. M., Green, D. K., and Allshire, R. C. (1990). Telomere reduction in human colorectal carcinoma and with ageing. *Nature*, **346**, 866-8.

Helms, C., Mishra, S. K., Riethman, H., Burgess, A. K., Ramachandra, S., Tierney, C., *et al.* (1992). Closure of a genetic linkage map of human chromosome 7q with centromere and telomere polymorphisms. *Genomics*, **14**, 1041-54.

Henderson, S. T. and Petes, T. D. (1992). Instability of simple sequence DNA in *Saccharomyces cerevisiae*. *Mol. Cell. Biol.*, **12**, 2749-57.

Henderson, S. T. and Petes, T. D. (1993). Instability of a plasmid-borne inverted repeat in *Saccharomyces cerevisiae*. *Genetics*, **133**, 57-62.

Henke, A., Fischer, C., and Rappold, G. A. (1993). Genetic map of the human pseudoautosomal region reveals a high rate of recombination in female meiosis at the Xp telomere. *Genomics*, **18**, 478-85.

IJdo, J. W., Baldini, A., Ward, D. C., Reeders, S. T., and Wells, R. A. (1991). Origin of human chromosome 2: an ancestral telomere-telomere fusion. *Proc. Natl. Acad. Sci. USA*, **88**, 9051-5.

IJdo, J. W., Lindsay, E. A., Wells, R. A., and Baldini, A. (1992a). Multiple variants in subtelomeric regions of normal karyotypes. *Genomics*, **14**, 1019-25.

IJdo, J. W., Baldini, A., Wells, R. A., Ward, D. C., and Reeders, S. T. (1992b). FRA2B is distinct from inverted telomere repeat arrays at 2q13. *Genomics*, **12**, 833-5.

Inglehearn, C. F. and Cooke, H. J. (1990). A VNTR immediately adjacent to the human pseudoautosomal telomere. *Nucleic Acids Res.*, **18**, 471-6.

Itzhaki, J. E., Barnett, M. A., MacCarthy, A. B., Buckle, V. J., Brown, W. R. A., and Porter, A. C. G. (1992). Targeted breakage of a human chromosome mediated by cloned human telomeric DNA. *Nature Genet.*, **2**, 283-7.

Jeffreys, A. J., Wilson, V., Kelly, R., Taylor, B. A., and Bulfield, G. (1987). Mouse DNA 'fingerprints': analysis of chromosome localization and germ-line stability of hypervariable loci in recombinant inbred strains. *Nucleic Acids Res.*, **15**, 2823-36.

Jeffreys, A. J., Neumann, R., and Wilson, V. (1990). Repeat unit sequence variation in minisatellites: a novel source of DNA polymorphism for studying variation and mutation by single molecule analysis. *Cell*, **60**, 473–85.

Julier, C., De Gouyon, B., Georges, M., Guénet, J. -L., Nakamura, Y., Avner, P., and Lathrop, G. M. (1990). Minisatellite linkage maps in the mouse by cross-hybridization with human probes containing tandem repeats. *Proc. Natl. Acad. Sci. USA*, **87**, 4585–9.

Kelly, R., Gibbs, M., Collick, A., and Jeffreys, A. J. (1991). Spontaneous mutation at the hypervariable mouse minisatellite locus *Ms6-hm*: flanking DNA sequence and analysis of germline and early somatic mutation events. *Proc. R. Soc. Lond. B*, **245**, 235–45.

Kipling, D. and Cooke, H. J. (1990). Hypervariable ultra-long telomeres in mice. *Nature*, **347**, 400–2.

Kipling, D., Ackford, H. E., Taylor, B. A., and Cooke, H. J. (1991). Mouse minor satellite DNA genetically maps to the centromere and is physically linked to the proximal telomere. *Genomics*, **11**, 235–41.

Knight, S. J. L., Flannery, A. V., Hirst, M. C., Campbell, L., Christodoulou, Z., Phelps, S. R., *et al.* (1993). Trinucleotide repeat amplification and hypermethylation of a CpG island in *FRAXE* mental retardation. *Cell*, **74**, 127–34.

Kremer, E. J., Pritchard, M., Lynch, M., Yu, S., Holman, K., Baker, E., *et al.* (1991). Mapping of DNA instability at the fragile X to a trinucleotide repeat sequence p(CCG)n. *Science*, **252**, 1711–14.

Kuwano, A., Ledbetter, S. A., Dobyns, W. B., Emanuel, B. S., and Ledbetter, D. H. (1991). Detection of deletions and cryptic translocations in Miller–Dieker syndrome by in situ hybridization. *Am. J. Human Genet.*, **49**, 707–14.

Lamb, J., Wilkie, A. O. M., Harris, P. C., Buckle, V. J., Lindenbaum, R. H., Barton, N. J., *et al.* (1989). Detection of breakpoints in submicroscopic chromosomal translocation, illustrating an important mechanism for genetic disease. *Lancet*, **ii**, 819–24.

Lamb, J., Harris, P. C., Wilkie, A. O. M., Wood, W. G., Dauwerse, J. G., and Higgs, D. R. (1993). De novo truncation of chromosome 16p and healing with (TTAGGG)$_n$ in the α-thalassemia/mental retardation syndrome (ATR-16). *Am. J. Human Genet.*, **52**, 668–76.

Levinson, G. and Gutman, G. A. (1987). Slipped-strand mispairing: a major mechanism for DNA sequence evolution. *Mol. Biol. Evol.*, **4**, 203–21.

Lin, C. C., Meyne, J., Sasi, R., and Moyzis, R. K. (1990). Apparent lack of telomere sequences on double minute chromosomes. *Cancer Genet. Cytogenet.*, **48**, 271–4.

Mahtani, M. M. and Willard, H. F. (1993). A polymorphic X-linked tetranucleotide repeat locus displaying a high rate of new mutation: implications for mechanisms of mutation at short tandem repeat loci. *Human Mol. Genet.*, **2**, 431–7.

Mariat, D., De Gouyon, B., Julier, C., Lathrop, M., and Vergnaud, G. (1993). Genetic mapping through the use of synthetic tandem repeats in the mouse genome. *Mammal. Genome*, **431**, 135–40.

Meltzer, P. S., Guan, X. -Y., and Trent, J. M. (1993). Telomere capture stabilizes chromosome breakage. *Nature Genet.*, **4**, 252–5.

Meyne, J., Ratliff, R. L., and Moyzis, R. K. (1989). Conservation of the human telomere sequence (TTAGGG)$_n$ among vertebrates. *Proc. Natl. Acad. Sci. USA*, **86**, 7049–53.

Migeon, B. R., McGinniss, M. J., Antonarakis, S. E., Axelman, J., Stasiowski,

B. A., Youssoufian, H., *et al.* (1993). Severe hemophilia A in a female by cryptic translocation: order and orientation of factor VIII within Xq28. *Genomics*, **16**, 20–5.

Moens, P. B. and Pearlman, R. E. (1990). Telomere and centromere DNA are associated with the cores of meiotic prophase chromosomes. *Chromosoma*, **100**, 8–14.

Morral, N., Nunes, V., Casals, T., and Estivill, X. (1991). CA/GT microsatellite alleles within the cystic fibrosis transmembrane conductance regulator (CFTR) gene are not generated by unequal crossingover. *Genomics*, **10**, 692–8.

Moyzis, R. K., Buckingham, J. M., Cram, L. S., Dani, M., Deaven, L. L., Jones, M. D., *et al.* (1988). A highly conserved repetitive DNA sequence, $(TTAGGG)_n$, present at the telomeres of human chromosomes. *Proc. Natl. Acad. Sci. USA*, **85**, 6622–6.

Murray, A. W., Claus, T. E., and Szostak, J. W. (1988). Characterization of two telomeric DNA processing reactions in *Saccharomyces cerevisiae. Mol. Cell. Biol.*, **8**, 4642–50.

Nakamura, Y., Carlson, M., Krapcho, K., Kanamori, M., and White, R. (1988). New approach for isolation of VNTR markers. *Am. J. Human Genet.*, **43**, 854–9.

NIH/CEPH Collaborative Mapping Group (1992). A comprehensive genetic linkage map of the human genome. *Science*, **258**, 67–86.

Orr, H. T., Chung, M. -Y., Banfi, S., Kwiatkowski, T. J., Servadio, A., Beaudet, A. L., *et al.* (1993). Expansion of an unstable trinucleotide CAG repeat in spinocerebellar ataxia type 1. *Nature Genet.*, **4**, 221–6.

Overhauser, J., Bengtsson, U., McMahon, J., Ulm, J., Butler, M. G., Santiago, L., and Wasmuth, J. J. (1989). Prenatal diagnosis and carrier detection of a cryptic translocation by using DNA markers from the short arm of chromosome 5. *Am. J. Human Genet.*, **45**, 296–303.

Park, V. M., Gustashaw, K. M., and Wathen, T. M. (1992). The presence of interstitial telomeric sequences in constitutional chromosome abnormalities. *Am. J. Human Genet.*, **50**, 914–23.

Pavan, W. J., Hieter, P., and Reeves, R. H. (1990). Generation of deletion derivatives by targeted transformation of human-derived yeast artificial chromosomes. *Proc. Natl. Acad. Sci. USA*, **87**, 1300–4.

Petit, C., Levilliers, J., and Weissenbach, J. (1988). Physical mapping of the human pseudo-autosomal region; comparison with genetic linkage map. *EMBO J.*, **7**, 2369–76.

Pezzolo, A., Gimelli, G., Cohen, A., Lavaggetto, A., Romano, C., Fogu, G., and Zuffardi, O. (1993). Presence of telomeric and subtelomeric sequences at the fusion points of ring chromosomes indicates that the ring syndrome is caused by ring instability. *Human Genet.*, **92**, 23–7.

Rappold, G. A. and Lehrach, H. (1988). A long range restriction map of the pseudoautosomal region by partial digest PFGE analysis from the telomere. *Nucleic Acids Res.*, **16**, 5361–77.

Reeve, A., Norman, A., Sinclair, P., Whittington-Smith, R., Hamey, Y., Donnai, D., and Read, A. (1993). True telomeric translocation in a baby with the Prader–Willi phenotype. *Am. J. Med. Genet.*, **47**, 1–6.

Richards, E. J. and Ausubel, F. M. (1988). Isolation of a higher eukaryotic telomere from Arabidopsis thaliana. *Cell*, **53**, 127–36.

Riethman, H. C., Moyzis, R. K., Meyne, J., Burke, D. T., and Olson, M. V. (1989). Cloning human telomeric DNA fragments into *Saccharomyces cerevisiae* using a yeast-artificial-chromosome vector. *Proc. Natl. Acad. Sci. USA*, **86**, 6240–4.

Riethman, H. C., Spais, C., Buckingham, J., Grady, D., and Moyzis, R. K. (1993). Physical analysis of the terminal 240 kb of DNA from human chromosome 7q. *Genomics*, **17**, 25–32.

Rossi, E., Floridia, G., Casali, M., Danesino, C., Chiumello, G., Bernardi, F., *et al.* (1993). Types, stability, and phenotypic consequences of chromosome rearrangements leading to interstitial telomeric sequences. *J. Med. Genet.*, **30**, 926–31.

Rouyer, F., de la Chapelle, A., Andersson, M., and Weissenbach, J. (1990). An interspersed repeated sequence specific for human subtelomeric regions. *EMBO J.*, **9**, 505–14.

Royle, N. J., Clarkson, R. E., Wong, Z., and Jeffreys, A. J. (1988). Clustering of hypervariable minisatellites in the proterminal regions of human autosomes. *Genomics*, **3**, 352–60.

Royle, N. J., Hill, M. C., and Jeffreys, A. J. (1992). Isolation of telomere junction fragments by anchored polymerase chain reaction. *Proc. R. Soc. Lond. Biol.*, **247**, 57–67.

Ruskin, B. and Fink, G. R. (1993). Mutations in *POL1* increase the mitotic instability of tandem inverted repeats in *Saccharomyces cerevisiae*. *Genetics*, **133**, 43–56.

Sawyer, J. R., Rowe, R. A., Hassed, S. J., and Cunniff, C. (1993). High-resolution cytogenetic characterization of telomeric associations in ring chromosome 19. *Human Genet.*, **91**, 42–4.

Schlötterer, C. and Tautz, D. (1992). Slippage synthesis of simple sequence DNA. *Nucleic Acids Res.*, **20**, 211–15.

Shampay, J. and Blackburn, E. H. (1988). Generation of telomere-length heterogeneity in *Saccharomyces cerevisiae*. *Proc. Natl. Acad. Sci. USA*, **85**, 534–8.

Shippey, C. A., Layton, M., and Secker-Walker, L.M. (1990). Leukemia characterized by multiple sub-clones with unbalanced translocations involving different telomeric segments: case report and review of the literature. *Genes Chromosomes Cancer*, **2**, 14–7.

Silver, A. and Cox, R. (1993). Telomere-like DNA polymorphisms associated with genetic predisposition to acute myeloid leukemia in irradiated CBA mice. *Proc. Natl. Acad. Sci. USA*, **90**, 1407–10.

Southern, E. M. (1970). Base sequence and evolution of guinea-pig α-satellite DNA. *Nature*, **227**, 794–8.

Starling, J. A., Maule, J., Hastie, N. D., and Allshire, R. C. (1990). Extensive telomere repeat arrays in mouse are hypervariable. *Nucleic Acids Res.*, **18**, 6881–8.

Tanzi, R. E., Haines, J. L., Watkins, P. C., Stewart, G. D., Wallace, M. R., Hallewell, R., *et al.* (1988). Genetic linkage map of human chromosome 21. *Genomics*, **3**, 129–36.

Vergnaud, G., Marlat, D., Zoroastro, M., and Lauthier, V. (1991). Detection of single and multiple polymorphic loci by synthetic tandem repeats of short oligonucleotides. *Electrophoresis*, **12**, 134–40.

Vollrath, D., Davies, R. W., Connelly, C., and Hieter, P. (1988). Physical mapping of large DNA by chromosome fragmentation. *Proc. Natl. Acad. Sci. USA*, **85**, 6027–31.

Wang, D., Fang, H., Cantor, C. R., and Smith, C. L. (1992). A contiguous *Not*I restriction map of band q22.3 of human chromosome 21. *Proc. Natl. Acad. Sci. USA*, **89**, 3222-6.

Warren, A. C., Slaugenhaupt, S. A., Lewis, J. G., Chakravarti, A., and Antonarakis, S. E. (1989). A genetic linkage map of 17 markers on human chromosome 21. *Genomics*, **4**, 579-91.

Weber, B., Collins, C., Robbins, C., Magenis, R. E., Delaney, A. D., Gray, J. W., and Hayden, M. R. (1990). Characterization and organization of DNA sequences adjacent to the human telomere associated repeat (TTAGGG)$_n$. *Nucleic Acids Res.*, **18**, 3353-61.

Weber, B., Allen, L., Magenis, R. E., Goodfellow, P. J., Smith, L., and Hayden, M. R. (1991a). Intrachromosomal location of the telomeric repeat (TTAGGG)$_n$. *Mammal. Genome*, **1**, 211-16.

Weber, B., Allen, L., Magenis, R. E., and Hayden, M. R. (1991b). A low-copy repeat located in subtelomeric regions of 14 different human chromosomal termini. *Cytogenet. Cell Genet.*, **57**, 179-83.

Weber, J. L. and Wong, C. (1993). Mutation of human short tandem repeats. *Human Mol. Genet.*, **2**, 1123-8.

Wells, R. A., Green, P., and Reeders, S. T. (1989). Simultaneous genetic mapping of multiple human minisatellite sequences using DNA fingerprinting. *Genomics*, **5**, 761-72.

Wells, R. A., Germino, G. G., Krishna, S., Buckle, V. J., and Reeders, S. T. (1990). Telomere-related sequences at interstitial sites in the human genome. *Genomics*, **8**, 699-704.

Wilkie, A. O. M. (1993). Detection of cryptic chromosomal abnormalities in unexplained mental retardation: a general strategy using hypervariable subtelomeric DNA polymorphisms. *Am. J. Human Genet.*, **53**, 688-701.

Wilkie, A. O. M. and Higgs, D. R. (1992). An unusually large (CA)$_n$ repeat in the region of divergence between subtelomeric alleles of human chromosome 16p. *Genomics*, **13**, 81-8.

Wilkie, A. O. M., Buckle, V. J., Harris, P. C., Lamb, J., Barton, N. J., Reeders, S. T., et al. (1990). Clinical features and molecular analysis of the α thalassemia/mental retardation syndromes. I. Cases due to deletions involving chromosome band 16p13.3. *Am. J. Human Genet.*, **46**, 1112-26.

Wilkie, A. O. M., Higgs, D. R., Rack, K. A., Buckle, V. J., Spurr, N. K., Fischel-Ghodsian, N., et al. (1991). Stable length polymorphism of up to 260 kb at the tip of the short arm of human chromosome 16. *Cell*, **64**, 595-606.

Wolff, R. K., Nakamura, Y., and White, R. (1988). Molecular characterization of a spontaneously generated new allele at a VNTR locus: no exchange of flanking DNA sequence. *Genomics*, **3**, 347-51.

Wolff, R. K., Plaetke, R., Jeffreys, A. J., and White, R. (1989). Unequal crossing over between homologous chromosomes is not the major mechanism involved in the generation of new alleles at VNTR loci. *Genomics*, **5**, 382-4.

Youngman, S., Bates, G. P., Williams, S., McClatchey, A. I., Baxendale, S., Sedlacek, Z., et al. (1992). The telomeric 60 kb of chromosome arm 4p is homologous to telomeric regions on 13p, 15p, 21p, and 22p. *Genomics*, **14**, 350-6.

Yunis, J. J. and Prakash, O. (1982). The origin of man: a chromosomal pictorial legacy. *Science*, **215**, 1525-30.

Index

Bold numbers denote reference to illustrations

Printed in the United States
1141100001B/31